Cancer Informatics in the Post Genomic Era

Toward Information-Based Medicine

Cancer Treatment and Research
Steven T. Rosen, M.D., *Series Editor*

Arnold, A.J. (ed.): *Endocrine Neoplasms.* 1997. ISBN 0-7923-4354-9.
Pollock, R.E. (ed.): *Surgical Oncology.* 1997. ISBN 0-7923-9900-5.
Verweij, J., Pinedo, H.M., Suit, H.D. (eds): *Soft Tissue Sarcomas: Present Achievements and Future Prospects.* 1997. ISBN 0-7923-9913-7.
Walterhouse, D.O., Cohn, S. L. (eds): *Diagnostic and Therapeutic Advances in Pediatric Oncology.* 1997. ISBN 0-7923-9978-1.
Mittal, B.B., Purdy, J.A., Ang, K.K. (eds): *Radiation Therapy.* 1998. ISBN 0-7923-9981-1.
Foon, K.A., Muss, H.B. (eds): *Biological and Hormonal Therapies of Cancer.* 1998. ISBN 0-7923-9997-8.
Ozols, R.F. (ed.): *Gynecologic Oncology.* 1998. ISBN 0-7923-8070-3.
Noskin, G. A. (ed.): *Management of Infectious Complications in Cancer Patients.* 1998. ISBN 0-7923-8150-5.
Bennett, C. L. (ed.): *Cancer Policy.* 1998. ISBN 0-7923-8203-X.
Benson, A. B. (ed.): *Gastrointestinal Oncology.* 1998. ISBN 0-7923-8205-6.
Tallman, M.S., Gordon, L.I. (eds): *Diagnostic and Therapeutic Advances in Hematologic Malignancies.* 1998. ISBN 0-7923-8206-4.
von Gunten, C.F. (ed.): *Palliative Care and Rehabilitation of Cancer Patients.* 1999. ISBN 0-7923-8525-X
Burt, R.K., Brush, M.M. (eds): *Advances in Allogeneic Hematopoietic Stem Cell Transplantation.* 1999. ISBN 0-7923-7714-1.
Angelos, P. (ed.): *Ethical Issues in Cancer Patient Care* 2000. ISBN 0-7923-7726-5.
Gradishar, W.J., Wood, W.C. (eds): *Advances in Breast Cancer Management.* 2000. ISBN 0-7923-7890-3.
Sparano, J. A. (ed.): *HIV & HTLV-I Associated Malignancies.* 2001. ISBN 0-7923-7220-4.
Ettinger, D. S. (ed.): *Thoracic Oncology.* 2001. ISBN 0-7923-7248-4.
Bergan, R. C. (ed.): *Cancer Chemoprevention.* 2001. ISBN 0-7923-7259-X.
Raza, A., Mundle, S.D. (eds): *Myelodysplastic Syndromes & Secondary Acute Myelogenous Leukemia* 2001. ISBN: 0-7923-7396.
Talamonti, M. S. (ed.): *Liver Directed Therapy for Primary and Metastatic Liver Tumors.* 2001. ISBN 0-7923-7523-8.
Stack, M.S., Fishman, D.A. (eds): *Ovarian Cancer.* 2001. ISBN 0-7923-7530-0.
Bashey, A., Ball, E.D. (eds): *Non-Myeloablative Allogeneic Transplantation.* 2002. ISBN 0-7923-7646-3.
Leong, S. P.L. (ed.): *Atlas of Selective Sentinel Lymphadenectomy for Melanoma, Breast Cancer and Colon Cancer.* 2002. ISBN 1-4020-7013-6.
Andersson , B., Murray D. (eds): *Clinically Relevant Resistance in Cancer Chemotherapy.* 2002. ISBN 1-4020-7200-7.
Beam, C. (ed.): *Biostatistical Applications in Cancer Research.* 2002. ISBN 1-4020-7226-0.
Brockstein, B., Masters, G. (eds): *Head and Neck Cancer.* 2003. ISBN 1-4020-7336-4.
Frank, D.A. (ed.): *Signal Transduction in Cancer.* 2003. ISBN 1-4020-7340-2.
Figlin, R. A. (ed.): *Kidney Cancer.* 2003. ISBN 1-4020-7457-3.
Kirsch, M.; Black, P. McL. (ed.): *Angiogenesis in Brain Tumors.* 2003. ISBN 1-4020-7704-1.
Keller, E.T., Chung, L.W.K (eds): *The Biology of Skeletal Metastases.* 2004. ISBN 1-4020-7749-1.
Kumar, R. (ed.): *Molecular Targeting and Signal Transduction.* 2004. ISBN 1-4020-7822-6.
Verweij, J., Pinedo, H.M. (eds): *Targeting Treatment of Soft Tissue Sarcomas.* 2004. ISBN 1-4020-7808-0.
Finn, W.G., Peterson, L.C. (eds.): *Hematopathology in Oncology.* 2004. ISBN 1-4020-7919-2.
Farid, N. (ed.): *Molecular Basis of Thyroid Cancer.* 2004. ISBN 1-4020-8106-5.
Khleif, S. (ed.): *Tumor Immunology and Cancer Vaccines.* 2004. ISBN 1-4020-8119-7.
Balducci, L., Extermann, M. (eds): *Biological Basis of Geriatric Oncology.* 2004. ISBN
Abrey, L.E., Chamberlain, M.C., Engelhard, H.H. (eds): *Leptomeningeal Metastases.* 2005. ISBN 0-387-24198-1
Platanias, L.C. (ed.): *Cytokines and Cancer.* 2005. ISBN 0-387-24360-7.
Leong, S.P.L., Kitagawa, Y., Kitajima, M. (eds): *Selective Sentinel Lymphadenectomy for Human Solid Cancer.* 2005. ISBN 0-387-23603-1.
Small, Jr. W., Woloschak, G. (eds): *Radiation Toxicity: A Practical Guide.* 2005. ISBN 1-4020-8053-0.
Haefner, B., Dalgleish, A. (eds): *The Link Between Inflammation and Cancer.* 2006. ISBN 0-387-26282-2.
Leonard, J.P., Coleman, M. (eds): *Hodgkin's and Non-Hodgkin's Lymphoma.* 2006. ISBN 0-387-29345.
Leong, S.P.L. (ed): *Cancer Clinical Trials: Proactive Strategies.* 2006. ISBN 0-387-33224-3.
Meyers, C. (ed): *Aids-Associated Viral Oncogenesis.* 2007. ISBN 978-0-387-46804-4.
Ceelen, W.P. (ed): *Peritoneal Carcinomatosis: A Multidisciplinary Approach.* 2007. ISBN 978-0-387-48991-9.
Leong, S.P.L. (ed): *Cancer Metastasis and the Lymphovascular System: Basis for rational therapy.* 2007. ISBN 978-0-387-69218-0.
Raizer, J., Abrey, L.E. (eds): *Brain Metastases.* 2007. ISBN 978-0-387-69221-0.
Jurisica, I., Wigle, D.A., Wong, B. (eds): *Cancer Informatics in the Post Genomic Era.* 2007. ISBN 978-0-387-69320-0

Cancer Informatics in the Post Genomic Era
Toward Information-Based Medicine

edited by

Igor Jurisica, PhD
Ontario Cancer Institute, PMH/UHN
Toronto Medical Discovery Tower
Toronto, Ontario

Dennis A. Wigle, MD, PhD
Division of Thoracic Surgery, Mayo Clinic
Mayo Clinic Cancer Center
Rochester, Minnesota, USA

Bill Wong, BSc, MBA
Program Director
Information Management
IBM Toronto Laboratory
Markham, Ontario

Igor Jurisica, PhD
Ontario Cancer Institute, PMH/UHN
Toronto Medical Discovery Tower
Division of Signaling Biology
Life Sciences Discovery Centre
Room 9-305, 101 College Street
Toronto, Ontario M5G 1L7 CANADA

Dennis A. Wigle, MD
Division of Thoracic Surgery
Mayo Clinic Cancer Center
200 First St. SW
Rochester, Minnesota 55905 USA

Bill Wong
Database Competitive Technologies
IBM Toronto Laboratory
8200 Warden Avenue
Markham, ON L3R 9Z7   CANADA

Series Editor:
Steven T. Rosen
Robert H. Lurie Comprehensive Cancer Center
Northwestern University
Chicago, IL
USA

Cancer Informatics in the Post Genomic Era: Toward Information-Based Medicine

Library of Congress Control Number: 2006939420

ISBN-10: 0-387-69320-3   e-ISBN-10: 0-387-69321-1
ISBN-13: 978-0-387-69320-0   e-ISBN-13: 978-0-387-69321-7

Printed on acid-free paper.

© 2007 Springer Science+Business Media, LLC
All rights reserved. This work may not be translated or copied in whole or in part without the written permission of the publisher (Springer Science+Business Media, LLC, 233 Spring Street, New York, NY 10013, USA), except for brief excerpts in connection with reviews or scholarly analysis. Use in connection with any form of information storage and retrieval, electronic adaptation, computer software, or by similar or dissimilar methodology now known or hereafter developed is forbidden.
The use in this publication of trade names, trademarks, service marks, and similar terms, even if they are not identified as such, is not to be taken as an expression of opinion as to whether or not they are subject to proprietary rights.

9 8 7 6 5 4 3 2 1

springer.com

Table of Contents

List of Figures vii
Foreword.. xi
Preface xiii
Contributing Authors xv
Acknowledgments.. xxi

Part I
Introduction 1
Dennis A. Wigle and Igor Jurisica

Part II
Bio-Medical Platforms 15
Ming Tsao

In Vivo Systems for Studying Cancer 25
Dennis A. Wigle, Jiang Liu, and Michael Johnston

Molecular Subtypes of Cancer from Gene Expression Profiling 45
Dennis A. Wigle and Igor Jurisica

Mass Spectrometry-based Systems Biology 59
Thomas Kislinger

Part III
Computational Platforms.. 85
Bill Wong and Igor Jurisica

Informatics 87
Bill Wong

Integrative Computational Biology 129
Igor Jurisica

Part IV
Future Steps and Challenges 147
Igor Jurisica and Dennis A. Wigle

Glossary.. 151
References 159
Index 179

List of Figures

Figure 1 Cancer research growth and utilization of high-throughput multiple platforms as indicated by number of PubMed references. The apparent decline in 2006 could be explained by not finalized numbers in PubMed. 4

Figure 2 EGFR—ras—MAPK signaling pathway from KEGG database (Ogata, Goto et al. 1999). 9

Figure 3 EGFR—ras protein interaction network from OPHID, visualized in NAViGaTor ver. 1.1 (http://ophid.utoronto.ca/navigator). Nodes in the graph are proteins, while edges correspond to interactions. Although EGFR, hras, kras and p53 are not directly linked, these major hubs in the network are highly mutually interconnected. 10

Figure 4 The variations between multiple samplings is significantly greater than those of elapsed time between sampling/freezing (Reprinted with permission from *Neoplasia*; (Blackhall, Pintilie et al. 2004)) 19

Figure 5 Performing frozen sections. Standard procedure for snap-frozen tissue bank sampling. 21

Figure 6 Xenograft tumors formed by established lung adenocarcinoma cell lines (A) and by primary xenograft lines (B). 23

Figure 7 a). A549 human lung adenocarcinoma cells implanted subcutaneously in nude mice. b). Individual dissected tumors. Courtesy Dr. Amy Tang, Mayo Clinic Cancer Center. 32

Figure 8 Thoracic cavity of a nude rat containing right caudal lobe tumor arising from NCI-H460 tumor fragments implanted endobronchially. Regional metastases to the mediastinal lymph nodes and systemic metastases to ribs and the left lung are shown. 35

Figure 9 TNM staging system for NSCLC. Reprinted with permission from (Wigle, Keshavjee et al. 2005). 46

Figure 10 Molecular profiles for SQCC 2B samples from (Wigle, Jurisica et al. 2002), visualized using self-organizing maps (SOMs) (Kohonen 1995) in BTSVQ clustering program (Sultan, Wigle et al. 2002). The first map shows a generalized gene expression patters, mapped into a color scheme. Each other map shows representation of one sample, clearly the first two samples being different from the last three samples. 48

Figure 11 Molecular profiling of stage I, II, III groups of NSCLC samples from (Wigle, Jurisica et al. 2002), using self-organizing maps (SOMs) (Sultan, Wigle et al. 2002). The heat maps clearly show both pattern similar within the stage – but also across stages. Importantly, the across stages patterns correlate with survival. 49

Figure 12 Multidimensional protein identification technology (MudPIT). (A) Complex protein mixtures are digested to peptides which are loaded onto microcapillary columns containing two chromatography resins. (B) Columns are place in-line with a HPLC pump and directly eluted into the mass spectrometer. Generated spectra are searched on computer clusters. (C) Shown is the basic concept of multi-step MudPIT runs. In each step a "salt bump" is used to move a sub-set of peptide bound to the SCX onto the RP resin. These peptides are then chromatographically separated and directly eluted into the MS. In the next step the salt concentration is increased to move another set of bound peptides from the SCX resin onto the RP resin. 64

Figure 13 Protein identification by mass spectrometry. (A) Proteins are separated by one-dimensional gel electrophoresis and bands of interest are excised from the gel and in-gel digested. The generated peptides are analyzed by MALDI-TOF-MS to generate a peptide mass fingerprint (PMF). (B) Protein identification by tandem mass spectrometry. First, the m/z of parent ions is recorded. Then individual peptide ions are isolated and fragmented by collision induced dissociation. Cross-correlation of theoretical MS/MS spectra generated by the search algorithm based on the mass of the parent ion with the experimental tandem mass spectra is used to identify the peptide sequence........................... 69

Figure 14 Multiple standards related to research, clinical trials and healthcare in general. Source: IBM.. 99

Figure 15 Trend of moving from current healthcare standards toward translational and personalized medicine by integrating information, and automating the diagnostic process. Source: IBM. 104

Figure 16 BlueGene scalability. BlueGene/Light integrates both computing and networking functions onto a single chip. The high level of integration and low power consumption permits dense packaging – 1,024 nodes (representing up to 5.6 Teraflops) can be housed in a single rack, and 64 racks can be integrated into one system. Source: IBM. 112

Figure 17 Growth of computational needs in biomedical field, as compared to the Moore's Law 114

Figure 18 OPHID protein-protein interaction web resource. http://ophid.utoronto.ca. Interactions can be searched in a batch mode using multiple identifiers (SwissProt, Unigene, Locuslink, etc.). The results are displayed in html, ASCII-delimited or PSI (Hermjakob, Montecchi-Palazzi et al. 2004) formats, or graphically, using NAViGaTor (http://ophid.utoronto.ca/navigator)................................ 117

Figure 19 Automated validation of predicted interaction using text mining of PubMed abstracts (Otasek, Brown et al. 2006). 118

Figure 20 Middleware for life sciences: WebSphere Information Integrator. 119

Figure 21 OLAP – converts relational tables to multi-dimensional database. 121

Figure 22 OLAP schema.. 124

Figure 23 Similar sequences. CDC55 (index=21) and CDC5 (index=37) are shown to match with a match fraction of 0.94 126

Figure 24 Similar sequences. CDC13 (index=10) and CDC17 (index =37) have a match of 0.8823 and this particular results are important because the scale of these two graphs is different and yet, IntelligentMiner has been able to find the similar sequence............... 127

Figure 25 A typical node of BTSVQ algorithm: (a) (left) Quantized gene set, computed with SOM for all samples. (centre) Representation of gene expression of 38 samples for genes selected by vector quantization. (b) Genes selected by SOM are clustered by minimizing within cluster distance and maximizing intra cluster distance (Davuos Boulin distance measure). (c) (centre) Child one of the root node generated by partitive k-means algorithm, with k=2. The visual representation of SOM component planes show that genes with lower levels of expression were separated from that with relatively high expression values by the partitive k-means algorithm. (left) Genes selected by vector quantization (using SOM) for the child one generated by partitive k-means algorithm. (d) Component planes and genes for child two. (e) Plot of genes selected by BTSVQ algorithm for a node. 136

Figure 26 Pseudo-color correlation matrix clustering. a) Shows the original correlation data on target proteins. Since the targets were selected based on previous analysis and knowledge of involved pathways, targets nicely show the squares around the diagonal (it is a symmetric matrix, high positive correlation is dark red; negative correlation is blue). Importantly, there is a strong crosstalk among specific groups of proteins (rectangles off the diagonal). b) To systematically enable the interpretation of such results, the correlation matrix can be clustered to identify protein groups and inter group relationships. 137

Figure 27 Visualization of protein complex data from (Collins, Kemmeren et al. 2007). Color represents cliques (highly interconnected protein complexes). Alpha-blending is used to suppress detail of the rest of the network. Visualized in 3D mode in NAViGaTor (http://ophid.utoronto.ca/navigator). 139

Figure 28 Integrated analysis of protein-protein interaction and microarray data. (A) Original DDR related PPI data from Figure 2 in (Ho, Gruhler et al. 2002). (B) Example of BTSVQ (Sultan, Wigle et al. 2002) analysis of yeast microarray data from (Hughes, Marton et al. 2000). (C) Graphical display of direct and indirect interactions of Rad17 with all 1,120 related proteins. (D) A weighted PPI graph that combines results from (A), (B), and (C) for Rad17. (E) A hypothesis generated from integrated PPI and microarray data involving PCNA-like complex from (A). 140

Figure 29 Integration of gene expression data with protein-protein interactions from OPHID (Brown and Jurisica 2005). The nodes in the network represent proteins; the color of the node represents annotated protein function when known (using GeneOntology). Lines connecting the nodes represent interactions between the two connected proteins. To emphasize interactions that are likely disrupted in cancer cells compared to non-malignant cells, in response to androgen, we use microarray data and co-expression between corresponding genes to "annotate" protein interactions. Black lines denote significantly correlated pairs in both groups, red lines denote correlation in cancer only, blue lines represent correlation in normal only, while dashed line represent no correlation. It clearly shows that there are full pathways and complexes that are only present in cancer samples (red lines). The highlighted (bold) line shows a known EGF pathway. Visualization done in NAViGaTor (http://ophid.utoronto.ca/navigator)...141

Foreword

The healthcare and pharmaceutical industries have been "buzzing" with the promise of personalized healthcare since the inception of the human genome project. Information technology will accelerate the delivery of advances in medical science and technology to the public. How will the convergence of information technology and life sciences impact the future?

During the past decade, life sciences and information technology began to converge, resulting in significant and life-impacting research – the result with perhaps the highest impact to date being the sequencing of the human genome and its influence on how clinical researchers now investigate methods and molecules that could improve the human condition. Knowledge gained through human genome sequencing is driving recent achievements in genomics, proteomics, molecular biology and bioinformatics. As the decade progresses, next generation medical science technology and capabilities, enabled by increasingly "smarter" information technology, will change the pace of discovery, development and delivery of new treatments even more dramatically. For example, biopharmaceutical research will continue to shift from a small, molecule-centered approach to one of stronger biomedical emphasis. This shift will focus on moving from the molecular actions of small molecule compounds toward delivering biological-based diagnostics and therapeutics. Healthcare will become increasingly personalized as these biological-based diagnostics and treatments become standard practice.

The application of information technology advances to those discoveries in science and medicine is giving rise to a new discipline, information-based medicine, which provides new knowledge by integrating and analyzing data from patients' clinical information, medical images, the environment, genetic profiles, as well as molecular and genomic research efforts. Information-based medicine is the marriage of information technology with the practice of medicine and pharmaceutical research for improved disease diagnosis, therapeutics and healthcare delivery. Information-based medicine is the use of information technology to achieve personalized medicine.

From developing health information networks for nations around the world to being a founding member of the Worldwide Biobank Summit, IBM is driving innovation in the healthcare and life sciences industries. IBM welcomes books like this that advances the industry's move toward information-based medicine and targeted treatment solutions.

Michael Svinte
Vice President of Information Based Medicine
IBM Healthcare and Life Sciences

Preface

Less than 50% of diagnosed cancers are cured using current treatment modalities. Many common cancers can already be fractionated into such therapeutic subsets with unique prognostic outcomes based on characteristic molecular phenotypes. It is widely expected that treatment approaches of complex cancer will soon be revolutionized by combining molecular profiling and computational analysis, which will result in the introduction of novel therapeutics and treatment decision algorithms that target the underlying molecular mechanisms of cancer.

The sequencing of the human genome was the first step in understanding the ways in which we are wired. However, this genetic blueprint provides only a "parts list", and neither information about how the human organism is actually working, nor insight into function or interactions among the ~30 thousand constitutive parts that comprise our genome. Considering that the 30 years of worldwide molecular biology efforts have only annotated about 10% of this gene set, and we know even less about proteins, it is comforting to know that high-throughput data generation and analysis is now widely available.

By arraying tens of thousands of genes and analyzing abundance of and interaction among proteins, it is now possible to measure the relative activity of genes and proteins in normal and diseased tissue. The technology and datasets of such profiling-based analyses will be described along with the mathematical challenges that face the mining of the resulting datasets. We describe the issues related to using this information in the clinical setting, and the future steps that will lead to drug design and development to cure complex diseases such as cancer.

Contributing Authors

Igor Jurisica, PhD

Dr. Jurisica is a Canada Research Chair in Integrative Computational Biology, a Scientist at the Ontario Cancer Institute, University Health Network since 2000, Associate Professor in the Departments of Computer Science and Medical Biophysics, University of Toronto, Adjunct Professor at School of Computing Science, Queen's University, and a Visiting Scientist at the IBM Centre for Advanced Studies. He earned his Dipl. Ing. degree in Computer Science and Engineering from the Slovak Technical University in 1991, M.Sc. and Ph.D. in Computer Science from the University of Toronto in 1993 and 1998 respectively.

Dr. Jurisica's research focuses on computational biology, and representation, analysis and visualization of high dimensional data generated by high-throughput biology experiments. Of particular interest is the use of comparative analysis for the mining of integrated datasets such as protein—protein interaction, gene expression profiling, and high-throughput screens for protein crystallization.

Scientist
Ontario Cancer Institute, PMH/UHN
Toronto Medical Discovery Tower
Division of Signaling Biology
Life Sciences Discovery Centre
Room 9-305
101 College Street
Toronto, Ontario M5G 1L7

Tel./Fax: 416-581-7437
Email: juris@ai.utoronto.ca
URL: http://www.cs.utoronto.ca/~juris

Associate Professor
Departments of Computer Science and Medical Biophysics,
University of Toronto

Dennis A. Wigle, MD, PhD

Since August 2006, Dennis Wigle has been a clinician-scientist at the Mayo Clinic Cancer Center in Rochester Minnesota. He is a practicing thoracic surgeon with an interest in thoracic oncology. His laboratory investigates the genetic basis and molecular sequence of events underlying thoracic malignancies. He holds an MD from the University of Toronto and a PhD from the Department of Anatomy and Cell Biology at Queen's University in Kingston, Canada. His interests include the application of novel computational methods to the analysis of high-throughput data in cancer biology.

Dennis A. Wigle
Division of Thoracic Surgery, Mayo Clinic
Mayo Clinic Cancer Center
200 First St. SW Rochester, Minnesota USA 55905

Tel: (507) 284-4099 (clinical office)
Tel: (507) 284-8462 (secretary)
Tel: (507) 538-0558 (lab office)
Fax: (507) 284-0058
Email: wigle.dennis@mayo.edu

Bill Wong

Bill Wong has an extensive background is software deployment technologies and has been working with a variety of database technologies. Some of his previous roles included being the Information Management product manager for Life Sciences, Linux, and Grid solutions. His current role is Program Director for Advanced Database Technologies at IBM. He works out of the

Toronto Lab and can often be found speaking at conferences on information management future trends and directions.

Program Director
Database Competitive Technologies
IBM Toronto Laboratory
8200 Warden Avenue
Markham, ON L3R 9Z7

Tel.: 905 413-2779, Fax: 905 413- 4928 T/L: (969)
Email: billw@ca.ibm.com

Thomas Kislinger, PhD

Dr. Kislinger is a Canada Research Chair in Proteomics in Cancer Biology, a Scientist at the Ontario Cancer Institute, University Health Network since 2006 and an Assistant Professor in the Department of Medical Biophysics at the University of Toronto. He earned his M.Sc. equivalent in Analytical Chemistry from the Ludwig-Maximilians University in Munich, Germany (1998) and his Ph.D. in Analytical Chemistry from the Friedrich-Alexander University in Erlangen, Germany and the Columbia University, New York (2001). He carried out his post-doctoral research at the Banting & Best Department of Medical Research in Toronto where he developed an expertise in large-scale expression proteomics of mammalian model systems.

Dr. Kislinger's research interests are focused on the development and application of shot-gun proteomics to diverse question in cancer, vascular and cardiovascular biology.

Scientist
Ontario Cancer Institute
MaRS Centre
Toronto Medical Discovery Tower
9th floor Room 9-807
101 College Street
Toronto, Ontario
Canada M5G 1L7

Telephone: 416-581-7627
Fax: 416-581-7629
e-mail: thomas.kislinger@utoronto.ca
URL: http://medbio.utoronto.ca/faculty/kislinger.html

Assistant Professor
Department of Medical Biophysics, University of Toronto

Ming-Sound Tsao, MD

Dr. Tsao is the M. Qasim Choksi Chair in Lung Cancer Translational Research and Professor of Laboratory Medicine and Pathobiology at University of Toronto. He is a Senior Scientist and Surgical Pathologist with special interest in neoplastic diseases of the aerodigestive tract. His focus on lung cancer research is on the identification and validation of molecular prognostic markers for early stage lung cancer patients, especially using genome-wide expression and genomic microarray platforms. He is also interested in predictive markers for benefits from adjuvant chemotherapy and targeted therapy in lung cancer. He has published more than 160 peer-reviewed manuscripts, with the most recent one published in the New England Journal of Medicine on molecular and clinical predictors of outcome in lung cancer patients treated by erlotinib.

Senior Scientist
Ontario Cancer Institute
Princess Margaret Hospital
610 University Avenue
Toronto Ontario M5G 2M9 Canada

Tel.: 416-340-4737
Email: ming.tsao@uhn.on.ca

Professor
Department of Laboratory Medicine and Pathobiology,
Univ. of Toronto

Chunlao Tang, PhD

Dr. Tang earned his PhD from the Molecular and Computational Biology Program at the University of Southern California. His primary research interest lies in studying the genetic basis underlying natural phenotypic variation. He seeks to elucidate the genetic variation associated with susceptibility to common diseases in humans using genomic approaches.

Email: chunlaot@usc.edu

Acknowledgments

Igor Jurisica gratefully acknowledges his lab members who contributed to the re-search results and stimulating research environment, especially Kevin Brown, Baiju Devani, Michael McGuffin, David Otasek, Mahima Agochiya, Dan Strumpf, Frederic Breard, Richard Lu and other students and programmers. Spe-cial thanks to numerous collaborators in lung, ovarian, prostate cancer sites.

Bill Wong would like to make the acknowledgements to the following people from IBM for their contributions: Mike Svinte and the Healthcare and life Sci-ences marketing staff, the IBM Business Consulting Services authors of the Per-sonalized Healthcare 2010 - Are you ready for information-based medicine paper, Barbara Eckman and Douglas Del Prete for the SQL queries, Richard Hale for his contributions regarding online analytical processing and data mining.

Dennis Wigle would like to thank lab members and members of the research community at Mayo Clinic who continue to inspire innovative approaches to problems in thoracic oncology.

Part I – Introduction

Dennis A. Wigle and Igor Jurisica

The sequencing of the human genome was widely anticipated for the contributions it would make toward understanding human evolution, the causation of disease, and the interplay between the environment and heredity in defining the human condition (Venter, Adams et al. 2001). The subsequent expected pace of discovery and its translation into benefit for the clinical management of cancer patients has not yet come to fruition. The expected landslide of genomic-based technologies for the molecular detection and diagnosis of cancer have yet to be clinically applied. Our fundamental understanding of the biology of cancer remains poor. Other than for a handful of notable exceptions, the rate of development and application of novel therapeutics has not appreciably changed in the post-genomic era.

Despite these facts, dramatic changes in clinical cancer management are beginning to appear on the horizon as a consequence of human genome sequencing and the technology development associated with the project. Molecular substaging for many tumor types is approaching clinical reality. Information from mutation analysis of specific genes is being incorporated into clinical decision making regarding chemotherapeutic agents. The pipeline of novel chemotherapeutics is full of promising new classes of agents with the potential for use in a patient-specific manner based on molecular substaging. It is an exciting time for translational and clinical cancer research.

However, as our understanding of cancer and its clinical treatment becomes ever more complicated, we have become burdened by the fact that data and knowledge management has become a significant hurdle to ongoing progress. The technological capacity to perform repeated biological and clinical observations to an exponentially greater degree, even more than previously thought possible, is both an exhilarating and frustrating experience. Managing the resulting information, even when focused on specific tumor types, has become a significant bottleneck.

Now that we are firmly in the post-genomic era of cancer care, we sought with this book to address a number of the issues related to the broad field of cancer informatics, where the bottlenecks are, and to discuss solution options as we go forward.

Understanding cancer biology – Cancer as a system failure

Decades of focused cancer research have demonstrated the oncogenic process to be frustratingly complex. Despite many triumphs in scientific and clinical understanding, we still do not comprehend the formation of most solid tumors at a basic level. This has hampered improvements in detection, diagnosis, and treatment strategies.

In our attempts to understand by reductionism, much work has gone into the biologic processes broadly described as the "hallmarks" of cancer. These include many diverse and seemingly nonoverlapping biological processes, including cell division, angiogenesis, migration and adhesion, DNA repair, and intracellular signaling (Hanahan and Weinberg 2000). Although some cancer subtypes are defined by a single genetic alteration leading to a primary defect in one of the above listed processes, most solid tumors responsible for the largest burden of human illness are heterogeneous lesions characterized by many if not all defects observable simultaneously. This includes lung, breast, prostate, colon, and central nervous system tumors among others. The integration of these observations are revealing that a true understanding of cancer biology will require a "systems" approach; an attempt to understand by viewing the hallmarks of cancer as an integrated whole rather than isolated, non-overlapping defects.

Why has the buzzword "systems biology" received so much recent attention? In short, it is because the key first step of defining system structures has quickly advanced from fantasy to reality in the post-genomic era. The achievement of full genome sequencing projects in many organisms, including human, has defined the initial "parts list" encoded in the medium of hereditary information transfer, DNA. The technological development associated with these achievements has spawned the nascent fields of genomics, proteomics, and multiple "-omic" disciplines defined by their systematic, non-hypothesis driven approaches to biological experimentation.

The life sciences are undergoing a profound transformation at the threshold of what is widely regarded as the century of biology (Kafatos and Eisner 2004). From a collection of narrow, well defined, almost parochial disciplines, they are rapidly morphing into domains that span the realm of molecular structure and function through to the application of this knowledge to clinical medicine. The results of teams of individual specialists dedicated to specific biological goals are providing insight into system structures and function not conceivable a decade ago. System level understanding, the approach advocated in systems biology, requires a change in our notion of "what to look for" in biology (Kafatos and Eisner 2004). While an understanding of individual genes and proteins continues to be important, the focus is superseded by the goal of understanding a systems structure, function and dynamics. System-level approaches and experimentation are computationally heavy and require a step back from the reductionist viewpoint that has dominated cancer research to date. Clearly the development of novel experimental models will be critical as we go forward to allow such approaches to be successful.

As can be seen in Figure 1, microarray technology, mass spectrometry, systems biology, and informatics approaches in cancer research are expanding exponentially. It is also apparent that experiments utilizing systems and informatics approaches for prediction and system modeling are not yet catching up to the volume of array profiling studies.

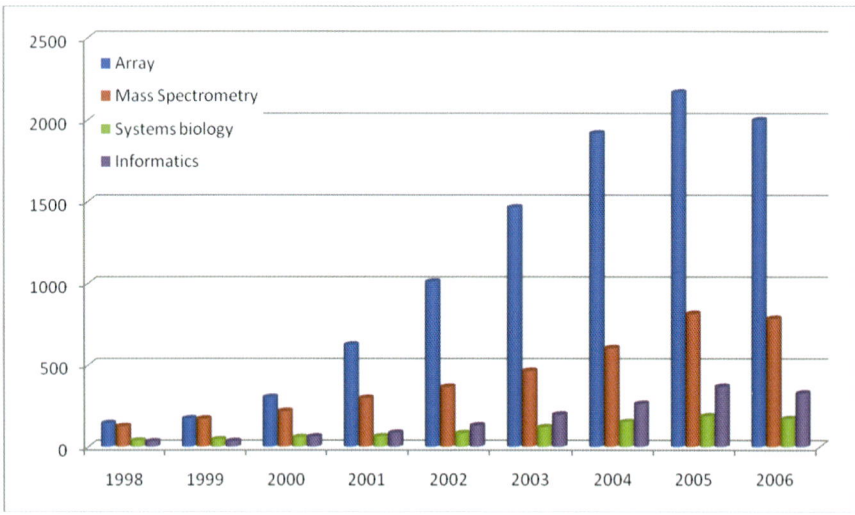

Figure 1. Cancer research growth and utilization of high-throughput multiple platforms as indicated by number of PubMed references. The apparent decline in 2006 could be explained by not finalized numbers in PubMed.

Molecular substaging: the road to a personalized medicine for cancer care

Much of the success of cancer treatment in the modern era rests on the ability to classify or stage patients to determine appropriate management strategy. Despite promising evidence that this may change in the near future for a number different cancers, the TNM classification systems for breast and prostate cancer are the only current cancers for which molecular information is utilized. No other solid tumor type has yet incorporated such molecular descriptors into the formal TNM staging system.

Early observations from many tumors demonstrated the potential for biologic classification of tumors into subgroups based on correlation with clinical outcome. This has been shown in preliminary data now from many tumor types, in some cases with molecular subtypes transcending traditional TNM stage classifications. The potential for molecular based staging to provide greater information than that available through current TNM systems has been a powerful driver for ongoing work in this area. Despite this promise however, clinically validated biomarker profiles are only now beginning to be tested in large patient cohorts to assess their translational utility. Using breast cancer as an example, gene-expression-profiling studies of primary breast tumors performed by different laboratories have resulted in the identification of a number of distinct prognostic profiles; however, many of these have little overlap in terms of gene identity. The earliest gene-expression profile test marketed in the United States is for early stage breast cancer. The Oncotype DX is a laboratory test that can be used on preserved (formalin-fixed, paraffin-embedded) stage I or II, estrogen receptor positive breast cancer tumor specimens from women whose tumors have not spread to their axillary nodes. Using the reverse transcription-polymerase chain reaction (RT-PCR), the test measures the level of expression of 21 specific genes to predict the probability of breast cancer recurrence. On the basis of those measurements, a "Recurrence Score" (RS) is assigned to the individual tumor. The lower the score, the lower the predicted probability of disease recurrence (Paik, Shak et al. 2004). Although this test is available for molecular diagnostic testing, it has not been validated prospectively in a clinical trial format.

Recent studies have suggested that the application of microarray technology for gene expression profiling of NSCLC specimens may permit the identification of specific molecular subtypes of the disease with different clinical behaviour (Bhattacharjee, Richards et al. 2001; Garber, Troyanskaya et al. 2001; Beer, Kardia et al. 2002; Wigle, Jurisica et al. 2002; Bild, Yao et al. 2006; Potti, Mukherjee et al. 2006). Data from individual studies however, although large by microarray standards, have not been of the magnitude required to make significant inferences about the relationships between gene expression and clinical parameters. A recent study in non-small cell lung cancer has demonstrated the potential utility of gene expression information in the clinical management of early stage lung cancer patients (Potti, Mukherjee et al. 2006). The advent of high-throughput platforms for the analysis of gene expression have provided the opportunity to look at potential correlations between gene expression biomarkers and clinical outcome. In NSCLC and other cancers, multiple publications have now demonstrated the possibility that clinically useful molecular substaging may be possible. Clinical translation of this technology is eagerly awaited by the oncology community.

Novel therapeutics: Genomics to drug discovery

The completion of the sequencing of the human genome, and those of other organisms, is expected to lead to many potential new drug targets in various diseases. It has long been predicted that novel therapeutic agents will be developed from high throughput approaches against such targets. The role of functional genomics in modern drug discovery is to prioritize these targets and to translate that knowledge into rational and reliable drug discovery. For the past several decades, drug discovery has focused primarily on a limited number of families of "druggable" genes against which medicinal chemists could readily develop compounds with a desired biochemical effect (Kramer and Cohen 2004). These targets were usually exhaustively investigated, with dozens or even hundreds of related publications often available, before huge investments in discovery programmes began. This has been altered in the post-genomic era. Although the genomics approach will undoubtedly increase the probability of developing novel therapies, the limited knowledge available for many putative targets has increased the risk and almost certainly the attrition rate for early-stage research projects.

To effectively exploit the information from human genome sequence, the incorporation of technologies capable of identifying, validating and prioritizing thousands of genes to select the most promising as targets will be required. The estimated 25,000 genes in the human genome, as well as multiple splice variants of many mRNAs, mandates that these technologies must be higher in throughput than most current technologies, as it will be impossible to develop the traditional depth of knowledge about each target. Importantly, no single technology will be sufficient to generate all of the necessary information, and the integration of knowledge from several approaches is required to select the best new drug targets for drug development (Kramer and Cohen 2004).

Modern pharmaceutical discovery is emerging as a new branch of science, thanks in large part to the technological advances that are allowing us to truly functionalize the genome. The investment made in sequencing the human and other genomes was made in reaction to the promise that this information would revolutionize medicine. In conjunction with the development of proteomic technologies based on mass spectrometry, the prospects for structure-based drug design are bright. However, integrating information across multiple technology platforms for the purpose of drug discovery represents an ever increasing challenge, as most currently available systems are not scalable to the task.

Information management in clinical oncology

The explosion in the number of compounds entering phase I trials has significantly increased the volume of clinical trial activity in modern oncology. Integration of the use of these novel compounds with established treatment regimens and new technologies in radiation therapy has expanded the therapeutic options for many cancer subtypes. Studying these permutations and combinations to determine effective doses and treatment regimen is a long and involved process. Despite this, the volume of trial activity has made it significantly more difficult to manage information for the modern clinical oncologist.

Data integration for biologic discoveries with potential therapeutic implications is an even greater problem. Making integrative portals for the non-computer scientist to visualize and interpret gene annotation, network information for signaling pathways, and place this in the context of clinical problems is an ongoing struggle. Solutions await for many of these issues.

Ultimately, the goal for medicine is to anticipate the need for medical treatment and define treatments that are specific for each person. Many coming developments will accelerate the pace of discovery by eliminating current unnecessary bottlenecks. The most prominent among these is the definition and deployment of a fully paperless medical record system for patient care. Although many institutions have made significant progress in this area, a limited few have achieved full implementation. The hoped-for electronic links between and among institutions will be dependent upon full utilization of such systems. Subsequent linkage to translational research databases harboring genomic, proteomic, and other information remains a further difficult but achievable task. It is however a necessary requirement if true personalized medicine in cancer care is ever to be achieved.

Case example: epidermal growth factor signaling networks in non-small cell lung cancer

Epidermal growth factor receptor (EGFR) was identified as a candidate for therapeutic control of cancer more than two decades ago. It is expressed in most patients with NSCLC, and has a role in cellular proliferation, inhibition of apoptosis, angiogenesis, metastatic potential, and chemoresistance (Blackhall, Ranson et al. 2006). The epidermal growth factor (EGF) receptor or EGFR belongs to the ErbB family, composed by four known cell membrane receptors with tyrosine kinase activity. The four members of the ErbB family are the EGFR (also known as ErbB-1/HER1), ErbB 2/Neu/HER2, ErbB-3/HER3, and ErbB-4/HER4. The molecular structure of each of these receptors is composed of an extra-cellular domain that recognizes and binds specific ligands, a trans-membrane domain, involved in interactions between receptors within the cell membrane, and an intra-cellular domain that contains the tyrosine kinase enzymatic activity (Ullrich and Schlessinger 1990). When activated, the tyrosine kinase domain catalyzes the phosphorylation of tyrosine residues on several intra-cellular signaling proteins, and on EGFR itself. The signaling pathway involves activation of ras, raf, and mitogen-activated protein kinase (MAPK), which determine the activation of several nuclear proteins that regulate cell cycle progression from G1 to S phase. Activation of the EGFR pathway is able to promote tumor growth and progression, stimulating cancer cell proliferation, production of angiogenic factors, invasion and metastasis, and inhibiting apoptosis.

Introduction 9

The protein-protein interaction network surrounding EGFR—ras—MAPK signaling contains a large number of well-characterized proteins, as shown on an example from KEGG database ((Ogata, Goto et al. 1999); http://www.genome.ad.jp/dbget-bin/show_pathway?hsa04010+1956 (Figure 2)). Information from protein interaction databases, such as OPHID (Brown and Jurisica 2005), further extends the potential to study and model these pathway under specific stimuli or in different tissues. Figure 3 shows one such visualization in NAViGaTor, highlighting only core proteins and suppressing other details by using alpha-blending. Individual nodes in the graph represent proteins, while edges correspond to known and predicted interactions. Color of nodes (except for red-highlighted ones) denotes different GeneOntology biological functions.

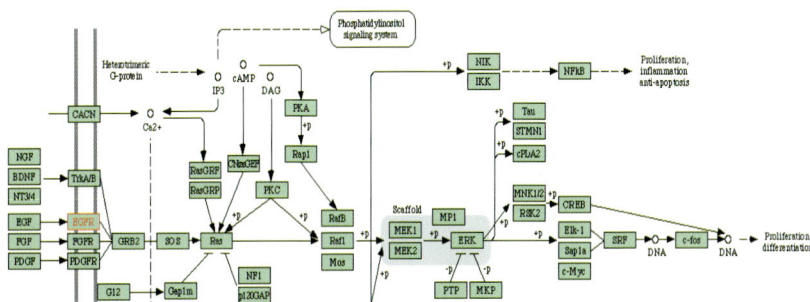

Figure 2. EGFR—ras—MAPK signaling pathway from KEGG database (Ogata, Goto et al. 1999).

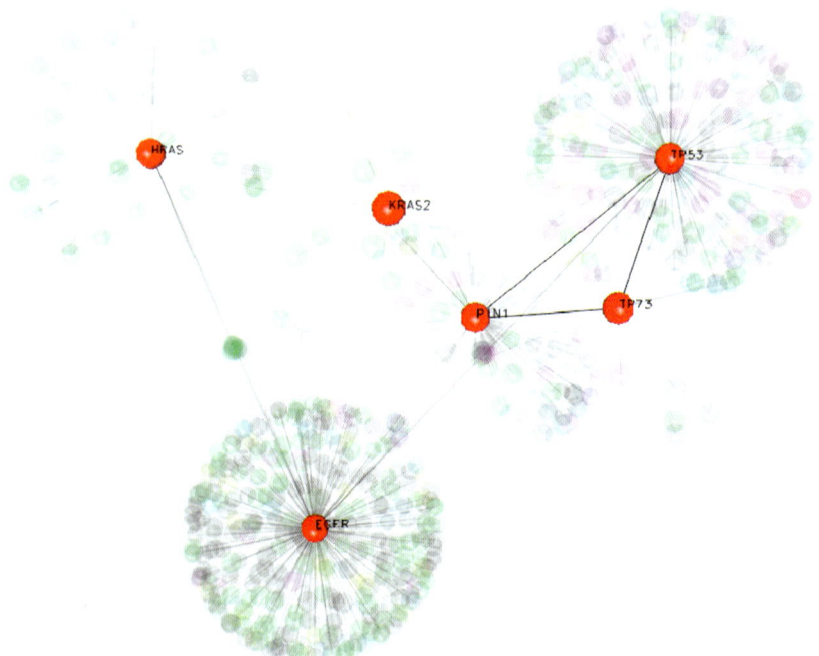

Figure 3. EGFR—ras protein interaction network from OPHID, visualized in NAViGaTor ver. 1.1 (**http://ophid.utoronto.ca/**navigator). Nodes in the graph are proteins, while edges correspond to interactions. Although EGFR, hras, Kras and p53 are not directly linked, these major hubs in the network are highly mutually interconnected.

In non-small cell lung cancer (NSCLC), the initial studies of epidermal growth factor receptor (EGFR) tyrosine kinase inhibitors (TKIs) brought significant enthusiasm for this targeted approach. Initial studies demonstrated that EGFR inhibition could lead to dramatic tumor regression in 10% to 15% of all treated patients. However, not all patients seemed to benefit from this treatment. A careful examination of patients who benefited from single-agent EGFR TKIs in phase II clinical trials, including unselected patients and those treated in the AstraZeneca gefitinib expanded access program, revealed clinical characteristics associated with an increased likelihood of a clinical or radiographic response. Patients most likely to achieve a radiographic response to EGFR TKIs were women, never smokers, patients with adenocarcinomas, and those of Japanese ethnicity.

In the spring of 2004, two simultaneously published studies examined case series of patients who had had dramatic clinical and/or radiographic responses to gefitinib (Lynch, Bell et al. 2004; Paez, Janne et al. 2004). Thirteen of 14 patients were found to have somatic activating mutations in the EGFR kinase domain, whereas none of the 11 patients who progressed on gefitinib had these EGFR mutations. Subsequently, EGFR mutations have been investigated in several series of NSCLC tumors from surgically resected patients and/or in patients treated with gefitinib or erlotinib. The mutation frequency appears to vary based on different patient characteristics, but very much mirrors the clinically defined subgroups deemed likely to achieve radiographic responses to EGFR TKIs. EGFR mutations are typically found in the first four exons of the tyrosine kinase domain of EGFR. Three types of mutations have been described: deletions in exon 19 account for about 60% of all mutations; a common missense mutation in exon 21 (L858R) accounts for another 25%; and, finally, rare point mutations in exons 18, 20, and 21 and insertion/duplications in exon 20 account for the remainder (Johnson and Janne 1257).

One of the startling aspects of the Paez et al. paper (Paez, Janne et al. 2004) was that despite sequencing of the exons encoding the activation loops of 47 of the 58 human receptor tyrosine kinase genes in the human genome in 58 NSCLC samples, only 3 of the tumors, all lung adenocarcinomas, showed heterozygous missense mutations in EGFR not present in the DNA from normal lung tissue from the same patients. No mutations were detected in amplicons from other receptor tyrosine kinase genes. All three tumors had the same EGFR mutation, predicted to change leucine-858 to arginine. Why EGFR is the sole RTK mutated in NSCLC is surprising and points to the important role of the receptor and its signaling axis.

A number of the early trials of EGFR-directed TKIs showed dissapointing results. The Iressa Survival Evaluation in advanced Lung cancer (ISEL) trial was designed to assess best supportive care with gefitinib or placebo in patients with NSCLC who had been treated previously (Thatcher, Chang et al. 2005). 1692 patients were enrolled from 210 centres in 28 countries across Europe, Asia, USA, South America, Australia, and Canada. The results showed a significantly higher objective response (i.e. complete response and partial response) for patients allocated gefitinib compared with those allocated placebo (8% vs 1%, $p < 0.0001$), but did not show a significant difference between groups in terms of survival. Preplanned subgroup analyses however did show a significant survival benefit for gefitinib in never smokers and in patients of Asian origin.

The results for the randomised, placebo-controlled phase III trial of erlotinib plus best supportive care (BR21; n = 731) by the National Cancer Institute of Canada were reported before those of ISEL (Shepherd, JR et al. 2005). In this study, median survival was 6.7 months and 1-year survival was 31% for patients treated with erlotinib, compared to 4.7 months and 22%, respectively, for placebo. Cox regression analysis of subgroups in BR21 showed higher survival in never smokers assigned gefitinib compared with those assigned placebo.

On the basis of data from BR21, erlotinib was approved by the FDA for patients with advanced, previously treated NSCLC in November, 2004. Consequently, when ISEL data became available in December, 2004, gefitinib was relabelled by the FDA for restricted use in patients already receiving it and obtaining a clinical benefit according to the view of the prescribing physician (Blackhall, Ranson et al. 2006).

In tumour samples from patients in the BR21 trial, *EGFR* mutations were associated with response to erlotinib but not with survival (Tsao, Sakurada et al. 2005). Assessment of *EGFR* mutations and clinical outcome for the ISEL trial is in progress. Several other molecular markers have been analysed for prediction of response to erlotinib or gefitinib. In particular, tumours with v-Ki-ras2 Kirsten rat sarcoma viral oncogene homologue (*KRAS*) mutations, which are common in NSCLC, might be resistant to EGFR tyrosine-kinase inhibitors (Jänne, Engelman et al. 2005), but results have not been reported for the predictive role of *KRAS* mutation status for response to an EGFR tyrosine-kinase inhibitor in the ISEL or BR21 trials (Blackhall, Ranson et al. 2006).

The EGFR TKI example demonstrates the potential for targeted agents directed in a personalized manner using molecular substaging. Clearly this is only the tip of the iceberg for targeted therapies in NSCLC. Integration of these with other agents targeting different pathways may herald the age of multitargeted small-molecule inhibitors that may come to supercede selective monotargeted agents (Blackhall, Ranson et al. 2006).

Summary

The issues associated with the development of EGFR-based TKIs demonstrate some of the challenges to drug discovery and its translation to the clinical management of cancer patients. These problems will intensify as more therapeutic targets are validated for potential intervention. Throughout the book we have attempted to illuminate some of the key issues related to information management and knowledge discovery in this new era of cancer care.

Part II – Bio-Medical Platforms

Ming Tsao

The pathology of human cancers is very complex. Tumors that develop in an organ or from a specific putative progenitor cell invariably consist of multiple types, which are currently best defined by their histological or cytological characteristics and/or clinical behavior. During the last two decades, increasing number of unique genetic abnormalities have been identified and associated with the tumors with specific clinical-pathological features (Vogelstein and Kinzler 2004). This has been most prominent for tumors of mesenchymal and hematopoietic cell origins, or those associated with hereditary syndromes. These discoveries have had significant impacts at the diagnostic and therapeutic levels, since these genetic abnormalities could represent the etiology and pathogenetic mechanisms for the development of these tumors.

The histopathology of most adult cancers is commonly heterogeneous. This is likely a phenotypic reflection of the diverse etiologies and complex genetic abnormalities that these tumors are associated with, most of which remain poorly defined. Nevertheless, there is a strong consensus that future and more effective cancer therapies are based on developing new drugs or therapeutic modalities that target the critical genetic or phenotypic aberrations occurring in the tumors (Arteaga, Khuri et al. 2002; Bild, Yao et al. 2006). Towards this goal, there is a general agreement among biomedical researchers that more precise definitions and classifications of human tumors based on their molecular genotypes and phenotypes are necessary. Molecular definition requires profiling at multiple levels, including at individual gene level (sequences, structure, copy number), expression level (mRNA and protein), as well as tissue organization and microenvironment level. The most basic requirement and at times the greatest barrier for accomplishing these works are the availability of good quality banked human tumor and the corresponding normal tissue.

Human Tissues Bank

There are many ways that human tissue and cells may be banked, as non-viable or viable tissues/cells. Non-viable tissues may be banked as chemically fixed or snap-frozen tissues. Viable tissue/cells may be banked by cryopreservation, as primary or propagable cell lines, or in the form of living xenograft tumors in immune deficient rodents. Each of these tissue-banking strategies has their respective advantages or disadvantages.

Paraffin embedded tissue bank

Throughout the world, there already exist in the Department of Pathology of every hospital, a very large bank of fixed human tissue representing all types of diseases. These paraffin embedded archival tissues are generally prepared using a standard histopathology protocol, as part of the routine surgical pathology practices. As legally and ethically required for good patient care practice, these blocks are commonly stored for 20 or more years, as required by the local health authorities. Since these tissue blocks are prepared for clinical diagnostic purposes, their processing and fixation protocol usually follows standard practices. In most instances, the protocol requires that tissues be placed immediately in a fixative, or as soon as possible after its resection or biopsy. In most instances, the fixative is a 10% buffered aqueous formaldehyde (formalin) solution. Formalin generates cross-links between proteins and nucleic acids (DNA and RNA), which results in their structural denaturation and fragmentation. This results in limitation for analyses by many quantitative techniques that require preservation of the full length and normal structure of the molecules being analyzed, such as RNA microarrays or proteomics analyses. However, formalin fixation and paraffin embedding also preserve the tissue, thus allowing them to be kept at low cost and in ambient temperature for many years.

The development of special techniques by microwave treatment to recover the antigenicity of formalin-denatured proteins has greatly enhanced the value of these materials for protein expression studies using the immunohistochemistry technique. The invention of tissue microarray (TMA) has further enhanced the value of paraffin tissue blocks in high throughput validation research on human tumors. In TMA, small (6-15 mm diameter) cores of formalin fixed and paraffin embedded tissue are arrayed into a single paraffin block. This allows the analysis and examination of a large number of tumor cases on a single histology slide and having been subjected to a specific stain. Recent improvement in the designs of microanalytical techniques for nucleic acids (quantitative polymerase chain reactions and microarrays) have also made it possible to perform global genomic and gene expression profiling experiments on paraffin embedded tissue materials.

Snap-frozen tissue bank

Until recently, many quantitative protein and nucleic acid studies that are performed on human tissue require fresh or snap-frozen banked samples. Despite recent improvements in the analytical techniques that allow greater scope of studies on formalin-fixed and paraffin embedded tissues, snap-frozen tissues remains the optimal materials for many studies. Despite this obvious importance of the quality of study samples, there is surprisingly a paucity of standardized protocols for the proper collection, processing and storage of human tissue samples for banking purposes.

Different types of molecules in tissue demonstrate various levels of stability. While RNA is notorious for rapid degradation by RNAse, the stability of RNA in biopsy or surgically resected tissues is largely undefined. Based on functional knowledge, it is expected that transcript encoding different classes of genes would demonstrate different half-lives, which would putatively influence their stability and decay rate after vascular devitalization. Blackhall et al. (Blackhall, Pintilie et al. 2004) investigated the stability of gene expression in surgically resected lung cancer for global expression pattern using cDNA microarray, and for the stability of stress and hypoxia related genes using the reverse transcription and quantitative PCR (RT-qPCR). Fragments of tissues were collected from lung tumors at various intervals up to 120 min after surgical resection. For some cases, several tissue fragments from different areas of the tumor were harvested at a single time point to study gene expression heterogeneity within the tumor. Each sample was snap-frozen after harvesting, and stored in liquid nitrogen until analysis. Remarkably, similar gene expression profiles were obtained for the majority of samples regardless of the time that had elapsed between resection and freezing. It was found that the variations between multiple samplings were significantly greater than those of elapsed time between sampling/freezing (Figure 4). The study concluded that tissue samples snap-frozen within 30-60 minutes of surgical resection are acceptable for gene expression studies, but sampling and pooling from multiple sites of each tumor appears desirable to overcome the molecular heterogeneity present in tumor specimens. Similar finding was reported by Hedley et al. (Hedley, Pintilie et al. 2003), who measured CA-IX in multiple biopsies using a semiautomated fluorescence image analysis technique and observed intratumoral heterogeneity to account for 41% of the variance in the data set.

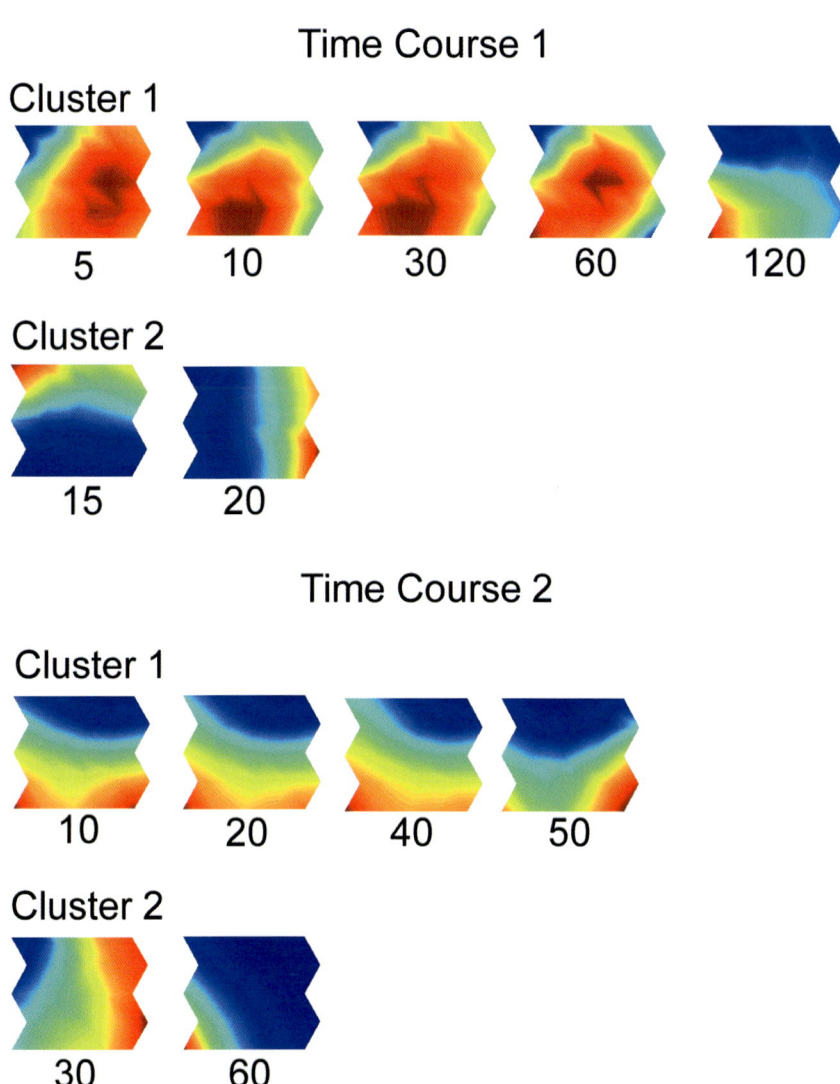

Figure 4. The variations between multiple samplings is significantly greater than those of elapsed time between sampling/freezing (Reprinted with permission from *Neoplasia*; (Blackhall, Pintilie et al. 2004)).

Most tumors are also composed of multiple cell types, including tumor cells, inflammatory cells, stroma fibro and myofibroblasts, and vascular endothelial cells. Heterogeneity in the composition of these cells may significantly influence the result of analysis performed. Therefore, it is imperative that each tissue that is subjected to molecular profiling study be rigorously quality controlled at histological level. This can be done in 2 ways: frozen section histology or formalin-fixed representative section histology.

Performing frozen sections on the study tissue allows a more accurate sampling of the cells or tissue to be analyzed. The latter can be enriched by microdissection from the stained frozen section slides. The disadvantage is that this is a very time consuming procedure that has to be performed by a very experienced person. The liability of thawed frozen tissue for rapid RNA degradation also represents a serious experimental risk. Frozen sections also do not provide the optimum histology for pathological evaluation of the tissue. Nevertheless, successful expression profiling of tumor tissue using this technique have been reported. An alternate method is to incorporate routinely during the tissue banking, sampling from the frozen tissue sample a representative tissue slice for formalin fixation and paraffin embedding (Figure 5). A regular histology section can then be from the tissue block for histopathological evaluation. An added advantage of this procedure is that the tissue in the paraffin block may also be used for immunohistochemistry studies that require rapid fixation of the tissue sample.

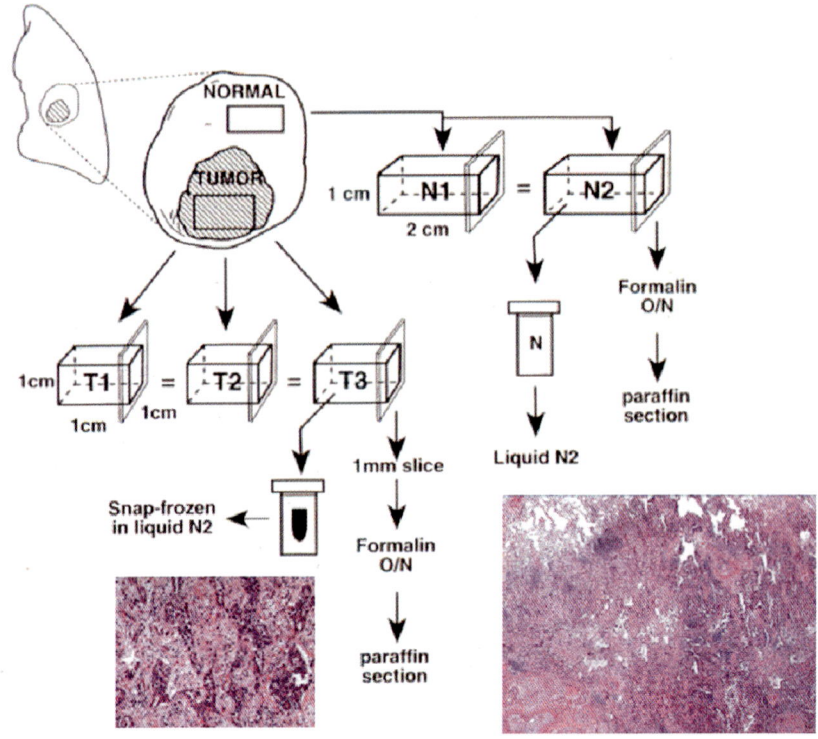

Figure 5. Performing frozen sections. Standard procedure for snap-frozen tissue bank sampling.

Non-frozen and non-chemically denatured tissue bank

Several other methods to preserve tissue in non-frozen condition and thus allowing the preservation of non-denatured molecules have also been tried. These include fixation in ethanol based chemicals or proprietary solutions, such as RNA*later*® (Ambion). The latter allows the isolation of intact RNA and DNA for profiling studies, but the suitability of tissue fixed in this solution for proteomics analysis is unknown.

Cultured tumor cell lines

Established human tumor cell lines represent the prototype of banking viable tumor cells. Through dedicated efforts of numerous investigators, a large number of propagable cell lines have been derived from most human tumor types or origin. These cell lines have played critical roles in our current understanding on the molecular aberrations and biology of human cancers. However, studies on cell lines present several drawbacks. The ability to establish cell lines from various types of human cancers is variable. Almost all small cell lung cancers when cultured may give rise to cell line. In contrast, only up to 25% of primary non-small cell lung cancer (NSCLC) cultures may lead to the establishment of cell lines. Cell lines appear easier to establish from advanced, poorly differentiated and metastatic cancers. The ability to establish cell line from the tumor has been reported to be a poor prognostic marker in NSCLC patients. Thus, although tumor cell lines demonstrate the genetic aberrations noted also in primary tumors, they may not be representative of the entire spectra of expression changes found in primary human tumors. Genome wide microarray studies have demonstrated that the expression profiles of cell lines tend to segregate separately from that of the primary tumors of same tumor type. However, the expression profiles of xenograft tumors from by these cell lines appear to recapitulate more closely that of the primary tumors.

Primary tumor xenograft tumor lines

Less available than cell lines, human primary tumor xenograft lines represent an alternate method of viable tissue bank. These lines were established by direct implantation of the primary human tumor tissue fragment into the subcutaneous or orthotopic sites of immune deficient mice. Unlike xenograft tumors formed by established cultured cell lines, the tumors formed by primary xenograft lines mostly preserve the histological phenotype of the primary tumors (Figure 6). Furthermore, the success rate of establishing xenograft tumor lines may be higher than that of establishing cultured tumor lines. The only drawback for setting up primary tumor xenograft tumor lines appear to be the higher cost of maintenance, and their less suitability for genetic manipulation that can be done easily in cultured cell lines.

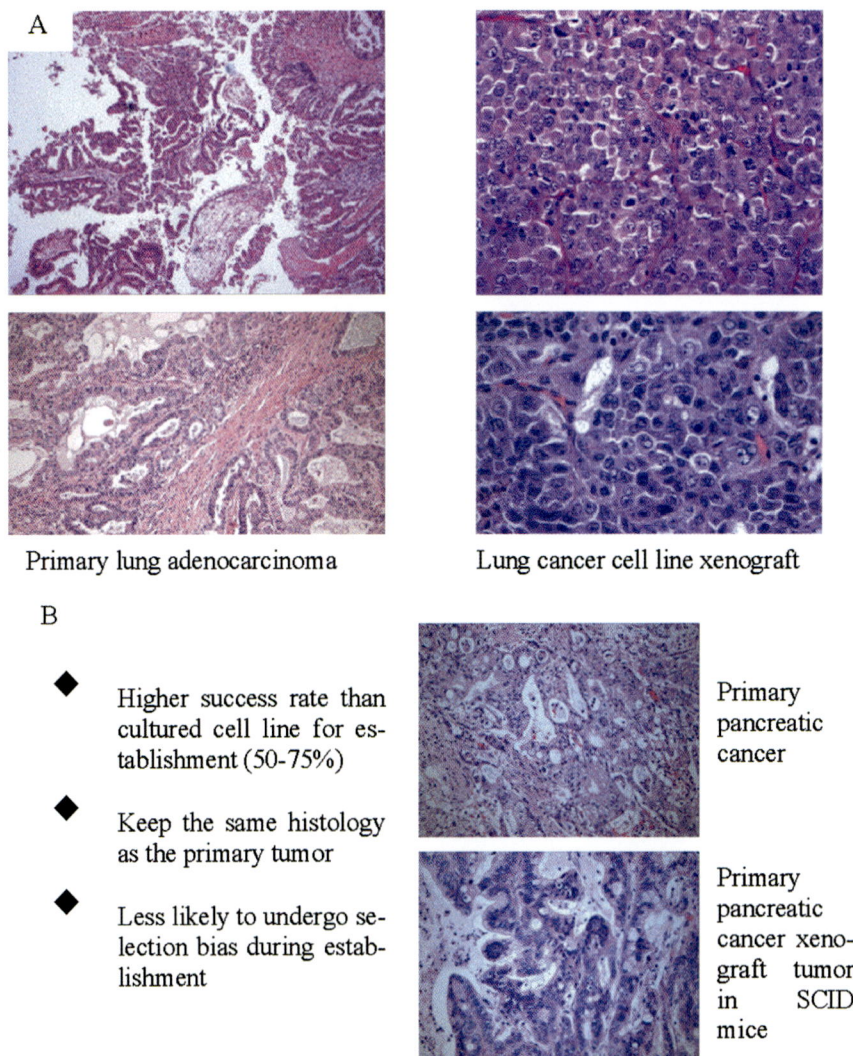

Primary lung adenocarcinoma   Lung cancer cell line xenograft

B

- ♦ Higher success rate than cultured cell line for establishment (50-75%)

- ♦ Keep the same histology as the primary tumor

- ♦ Less likely to undergo selection bias during establishment

Primary pancreatic cancer

Primary pancreatic cancer xenograft tumor in SCID mice

Figure 6. Xenograft tumors formed by established lung adenocarcinoma cell lines (A) and by primary xenograft lines (B).

In Vivo Systems for Studying Cancer

Dennis A. Wigle, Jiang Liu, and Michael Johnston

Introduction

Although the past few decades have seen great strides for cancer research, the molecular pathogenesis of most solid tumors from many tissue types remains largely undefined. Most of what we know about the molecular steps involved in cancer formation comes from defined genetic manipulations in the mouse and other model organisms. In lung cancer, the lack of defined models has hampered our understanding of disease progression and potential therapeutic strategies. Such models are essential tools to facilitate the development of new therapies.

Lung cancer continues to be the leading cause of cancer-related death worldwide (Kerr 2001). Despite aggressive local and systemic therapies (Johnston, Mullen et al. 2001), the majority of patients succumb to progressive metastatic disease. The defined molecular steps involved in the pathogenesis of lung cancer unfortunately remain elusive. Non-small cell lung cancer, a subset of lung cancer, is characterized by its aggressive biology and heterogeneity in clinical outcome. Humans are one of only a few species susceptible to developing spontaneous lung cancer. Lung tumors in domestic animals are periodically observed by veterinarians, but Livingood's histologic description 100 years ago of a papillary tumor in a mouse (Livingood 1986) initiated the idea of using animals as experimental model systems. Currently, several types of animal models have been developed for experimental lung cancer research. These include transgenic mouse models, chemically induced lung tumors, and human tumor xenografts.

The biology of cancer is rapidly emerging as one of the most difficult systems biology problems. The myriad of genetic alterations and their phenotypic outputs create an exceptionally complex picture to dissect from a reductionist viewpoint. Cancer models that accurately reflect these changes are difficult to generate. It is unlikely that a single model system can faithfully reflect the whole process of cancer development and progression, and as a consequence this requires us to interpret results from model systems with caution. However, appropriate use of available model systems, with an appropriate understanding of their limitations, provides a valuable and necessary tool to study malignant transformation. This chapter summarizes a number of lung cancer model systems in use as a context for the discussion of in vivo systems for studying cancer. We attempt to define both their utility and limitations.

General Principles

Many aspects of experimental cancer research require the use of animal model systems to reflect the true system context of oncogenesis *in vivo*. Tumor-host interactions including immunologic effects, vascular and stromal effects, and host-related pharmacologic and pharmacokinetic effects are examples poorly modeled *in vitro*. Several studies have shown that lung tumors developed in mice or rats are quite similar in histologic morphology and molecular characteristics to human lung cancer (Malkinson 1992; Howard, Mullen et al. 1999). In general, the spontaneous or chemically induced tumor models that are either idiopathic or arise following a carcinogenic stimulus (Corbett 1975; Corbett 1984) most closely mimic the true clinical situation. Unfortunately, these tumors are usually measurable only late in their course, their metastatic pattern is not uniform, and their response to therapy is generally poor.

Transplanted animal tumor models and human tumor xenografts are widely used in experimental therapeutics. Since malignant cells or tissue are directly inoculated into the host animal, effects on early events, such as initiation and carcinogenesis, are not suited for study with these models. However, tumor growth, invasion and metastasis are amenable for investigation, since tumor development uniformly follows inoculation with predictable growth and metastatic patterns. Testing of new therapeutic approaches and imaging strategies are particularly well suited for these models.

Transgenic technology has allowed for the development of mouse models for lung cancer. The mouse is the only genetically tractable model organism with lungs similar in structure and function to humans, and the only model organism that develops cancers of similar histopathologies to that seen clinically. The ability to target regulatory genes to the lungs in a cell-specific fashion is feasible with modern gene transfer technologies. These genetically engineered mouse lung cancer models can be exploited to define the molecular events that contribute to the pathogenesis of this disease.

Transgenic Lung Cancer Models

Transgenic mouse technology has proved extremely useful to create models of tumor development, cloning immortalized cellular subpopulations, and testing experimental therapeutic approaches (Adams and Cory 1991; Fowlis and Balmain 1993; Thomas and Balkwill 1995). The ability to integrate a gene of interest into the genome of an animal provides a novel approach for cancer investigation. Gene transfection can be achieved with microinjection (Gordon and Ruddle 1983; Brinster, Chen et al. 1985), retroviral infection, or embryonic stem cell transfer (Jaenisch 1980; Jähner and Jaenisch 1980; Jaenisch, Jahner et al. 1981; Soriano and Jaenisch 1986). Transgenic mice are excellent models for studying the consequences of oncogene expression in animals, the effect of oncogenes on growth and differentiation, and their potential for cellular transformation.

Transgenic mice also provide an *in vivo* preclinical model for gene therapy and gene transfer. An example of this technique as applied to drug development is the introduction of the multiple drug resistance (mdr-1) gene into transgenic animals (Galski, Sullivan et al. 1989). The mdr-1 gene, which is expressed in marrow stem cells, protects cancer cells from damage by extruding cytotoxic chemotherapeutic agents from the cell and confers *in vivo* resistance to drug toxicity in the whole animal. Such animal models have the potential for identifying agents, or combinations of agents, which are nontoxic to the animal but inhibit the function of the mdr-1 gene or its product and reversing the resistance phenotype.

Transgenic models capable of inducing lung cancer have also been developed. When mutated K-*ras*, p53 or SV40 T antigen are used as transgenes and integrated into the host genome, lung tumors develop in mice soon after birth and result in early death of the animal. These genes may be non-specifically expressed throughout the body or linked to lung-specific promoters so that their expression is selectively targeted to non-ciliated Clara cells or alveolar type II pneumocytes (Suda, Aizawa et al. 1987; Maronpot, Palmiter et al. 1991; Wikenheiser 1992; Sandmoller 1995). Although these animals have been used to a limited extent to investigate the molecular events involved in the progression of lung cancer, the rapid progression and early onset of cancer makes investigation of the early events involved in cancer development difficult (Zhao 2000).

The field of transgenic technology has now evolved to allow an investigator more control over specific transgenes. Bitransgenic systems are the most effective gene regulatory systems for transgenic mice, with the tetracycline-based regulatory system (Shockett and Schatz 1996) being the most commonly used. This system, which is under the control of elements responsive to tetracycline or its analogues, has at least two advantages over conventional transgenic mice. First, the transgene can, in principle, be turned on at any time, and thus resembles a somatic mutation. Second, regulated loss of expression (turning off the transgene) can be used to determine whether the transgene is required to maintain growth and proliferation of the tumor. A transgenic mouse model of lung adenocarcinoma with expression of a mutant active K-*ras* transgene has been developed by using this regulatory transgenic technology (Fisher, Wellen et al. 2001). Tumors rapidly regress as a result of apoptosis when doxycycline, a tetracycline analog, is withdrawn, demonstrating the role of K-*ras* in driving lung tumorigenesis. Several other lung cancer mouse models have also been developed with conditional activation of oncogenic K-*ras* (Jackson, Willis et al. 2001; Johnson, Mercer et al. 2001; Meuwissen, Linn et al. 2001). The use of regulatory transgenic systems such as these is a valuable tool to identify targets for future drug development.

One of the issues with a number of known oncogenes and/or tumor suppressors is that they are embryonic lethal when deleted in the mouse. As a consequence, the study of tissue specific pathways of tumorigenesis involving these genes is impossible. Although explored in only limited fashion to date, the potential of tissue specific deletions using the cre-loxP system has great potential for the dissection of tissue-specific tumor pathways. Many of these avenues remain to be explored.

Chemically Induced Lung Cancer Models

Humans are constantly exposed to potentially harmful mixtures of chemicals and physical agents from the environment. The laboratory environment allows controlled administration of such toxins to animals. Mice that are prone to develop spontaneous lung tumors are also often susceptible to chemically induced lung cancer (Jackson, Willis et al.). If a newborn inbred strain A/J mouse is given a single intraperitoneal injection of ethyl carbamate (urethane) at a dose of more than 0.5 mg/g body weight, it will develop dozens of benign lung adenomas within a few months (Shimkin and Stoner 1975). Some of these induced tumors eventually progress to adenocarcinomas that are histopathologically indistinguishable from human adenocarcinoma (Malkinson 1992). Many chemicals and environmental agents have been tested for carcinogenic activity using this tumorigenic response of the mouse lung as an indicator of toxicity.

Strain A mice have also been extensively used as a murine lung tumor bioassay to assess carcinogenic activity of chemicals, including urethane, benzopyrene, metals, aflatoxin, and constituents of tobacco smoke such as polyaromatic hydrocarbons and nitrosamines (Shimkin and Stoner 1975; Stoner 1991; Kim and Lee 1996). These agents can act as initiators and/or promoters of pulmonary tumorigenesis by accelerating tumor onset and increasing tumor multiplicity. In addition to chemicals, both radiation and viruses can induce lung tumors in mice (Rapp and Todaro 1980). Although induction of lung tumors in such models is highly reproducible (Malkinson 1989), all chemically induced lung tumors for some unknown reason exhibit relatively low metastatic potential.

Besides its use in carcinogen detection, the strain A mouse lung tumor model is employed extensively to identify inhibitors of chemical carcinogenesis. A number of chemopreventive agents, including beta-naphthoflavone (Anderson and Priest 1980), butylated hydroxyanisole (Wattenberg 1972), phenethyl isothiocyanate (Morse 1991), green tea and black tea (Wang, Fu et al. 1992) have been shown to inhibit chemical induced lung tumors in strain A mice. In most instances, inhibition of lung tumorigenesis has been correlated with effects of the chemopreventive agents on metabolic activation and/or detoxification of the respective carcinogen involved.

Various anti-inflammatory drugs can also inhibit mouse lung tumorigenesis. These include nonsteroidal anti-inflammatory drugs such as indomethacin, sulindac and aspirin (Jalbert and Castonguay 1992; Duperron and Castonguay 1997). Anti-inflammatory drugs that induce regression of benign colonic polyps in humans are modestly effective at lowering lung tumor incidence and multiplicity in mice (Duperron and Castonguay 1997). Interestingly, the density of apoptotic cell bodies is increased 3-fold in lung adenomas in A/J mice treated with indomethacin (Moody, Leyton et al. 2001). A new approach uses drugs that selectively inhibit the inducible Cox-2 enzyme associated with inflammation, without inhibiting the constitutive Cox-1 enzyme necessary for protecting digestive epithelial mucosa. Recent studies have revealed that the Cox-2 inhibitors JTE-522 and nimesulide can reduce regional lymph node and lung metastases in an *in vivo* lung cancer model (Kozaki, Koshikawa et al.).

Two hamster models have been used by the National Cancer Institute Chemoprevention Branch to evaluate efficacy against respiratory tract cancers. This includes MNU-induced tracheobroncheal squamous cell carcinomas and DEN-induced lung adenocarcinomas (Steele, Moon et al. 1994). In the DEN model (Moon, Rao et al. 1992) twice-weekly subcutaneous injections of 17.8 mg DEN/kg for 20 weeks starting at age 7-8 weeks produce lung tumors in 40-50%

of male Syrian hamsters. Serial studies demonstrate that most lung tumors originate from the respiratory Clara and endocrine cells (Schuller 1985). This model may be particularly appropriate for examining the chemopreventive activity of chemical agents in small cell lung cancer, a tumor originating from neuroendocrine cells in the lung.

Despite the usefulness of carcinogen-induced lung cancer models, major disadvantages remain. They are time consuming and, more importantly, they yield a variety of different histological tumor cell types with variable natural histories that might not be directly relevant to human lung cancer.

Human Lung Tumor Xenografts

The success of human tumor xenografting into immunocompromised rodents and the ability to maintain the histologic and biologic identity of tumor cells through successive passages *in vivo* revolutionized many aspects of cancer research, including drug development (Povlsen and Rygaard 1971). Since the immunogenicity of human neoplasms causes their destruction when implanted into immunocompetent species, experimental hosts need to be immunocompromised. Irradiation, thymectomy, splenectomy and corticosteroids were initially used to blunt acquired immunity. With the breeding of hairless nude mouse mutants (nu/nu homozygotes), severe combined immunodeficient (SCID) mice and Rowett nude rats, these laboratory animals have now become the most common recipients of human tumors in experimental therapeutics.

Subcutaneous implantation is the predominant site to transplant human tumor material into the nude mouse, since the procedure is simple and the site is readily accessible (Figure 7). This also allows for straightforward monitoring of tumor growth. Although subcutaneous xenograft models can predict clinical efficacy (Steel, Courtenay et al. 1983; Mattern, Bak et al. 1988; Boven 1992), these models have significant limitations, which include: (1) A low tumor take rate for fresh clinical specimens, with the percentage varying widely depending on the type of cancer (Mattern, Bak et al. 1988).

(2) The unusual tissue compartment where tumor growth occurs. This raises the question of how the microenvironment of the subcutaneous space might influence study results. (3) The lack of consistent invasion and metastasis is perhaps the greatest limitation of the

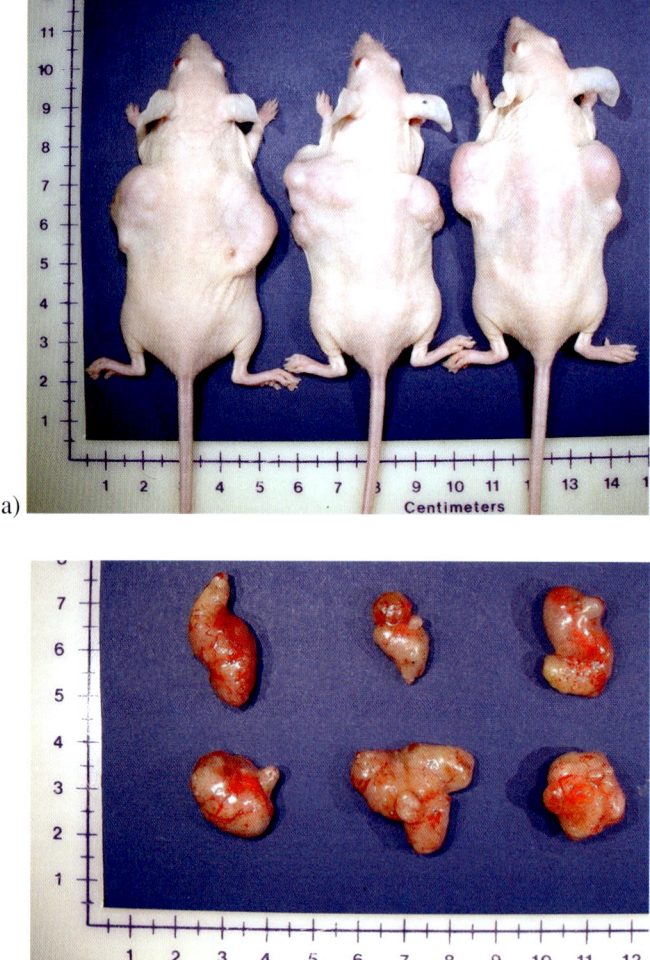

Figure 7. a) A549 human lung adenocarcinoma cells implanted subcutaneously in nude mice. **b)** Individual dissected tumors. Courtesy Dr. Amy Tang, Mayo Clinic Cancer Center.

model (Fidler 1986; Mattern, Bak et al. 1988), because these properties of cancer are most closely linked to clinical outcome. (4) Since tumor-bearing animals may succumb to local tumor effects, such as infection from skin ulceration, survival is not a feasible endpoint for assessing drug efficacy in these animals.

Because of the above limitations, orthotopic models were developed where human tumors are implanted directly into the appropriate organ or tissue of origin in the laboratory animal. The advantages of these models, such as improved tumor take and enhanced invasive and metastatic properties, are now well established (Fidler 1986; Fidler, Naito et al. 1990; Fidler 1991). Orthotopic implantation permits the expression of the metastatic phenotype of a variety of tumors; for example, colon carcinoma cells grown in the cecal wall, bladder carcinoma in the bladder, renal cell carcinoma cells under the renal capsule, and melanomas implanted subdermally, all yield metastases at much higher frequency than when grown subcutaneously (Kerbel, Cornil et al. 1991; Manzotti, Audisio et al. 1993). In contrast to subcutaneous implantation models, orthotopic models demonstrate that non-small cell lung cancer (NSCLC) cell lines implanted into the thoracic cavity of nude mice are almost always fatal (McLemore 1988). Orthotopic implantation appears to result in more aggressive tumor biology and shorter animal survival. This suggests that the local environment for in situ growth may reflect clinical reality more closely than heterotopic tumor implantation. An organ-specific site presumably provides tumor cells with the most appropriate milieu for studying local growth and metastasis. These manifestations support Paget's hypothesis that metastasis is not a random phenomenon. Rather, he concluded, malignant cells have special affinity for growth in the environment of certain organs, the familiar seed and soil theory (Paget 1889).

Orthotopic lung cancer models have been developed using endobronchial, intrathoracic or intravenous instillation of tumor cell suspensions (McLemore 1987; Howard 1991; Wang, Fu et al. 1992) and surgical implantation of fresh, histologically intact tumor tissue (An, Wang et al. 1996; Rashidi, Yang et al. 2000). The first orthotopic model of human lung cancer was developed by McLemore et al. (McLemore 1987) who implanted human lung cancer cell lines and enzymatically dissociated human lung tumors in the right lung of nude mice via an endobronchial injection. The tumors had increased growth and invasiveness within the lung as compared to the same tumors inoculated subcutaneously. However, most of the tumors remained within the right lung, with only 3% showing distant spread to lymph nodes, liver or spleen (McLemore 1987). McLemore et al. (McLemore 1988) subsequently developed a second model by percutaneously injecting lung tumor cells via an intrathoracic route into the pleural space. The model gave high tumor take-rates, with reproducible growth and a mortality endpoint as a result of local disease progression; however, very little metastases were observed. This approach appears to have major drawbacks, which may result in seeding lung cancer cells into the pleural space rather than within the pulmonary parenchyma or bronchi where clinical human lung cancer originates.

Similarly, endobronchial implantation has been used to grow non-small cell (A549, NCI-H460, and NCI-H125) and small cell (NCI-H345) lung carcinoma lines in irradiated nude rats (Howard 1991). In these models, metastases to the mediastinal lymph nodes are frequently seen, but the incidence of systemic metastasis is low. In order to further develop a model capable of metastasizing both regionally and systemically from a primary bronchial site, fresh tumor fragments derived from orthotopic lung tumors were implanted and grown from the H460 cell line. This H460 nude rat model has a 100% tumor take-rate in the lung with a rapid and reproducible growth rate up to approximately four grams over a 32-35 day period (Figure 8). It also metastasizes at a consistent rate to both regional mediastinal lymph nodes and distant systemic sites, including bone, brain and kidney. This is the first human lung cancer model to show extensive systemic metastasis from an orthotopic primary site (Howard, Mullen et al. 1999).

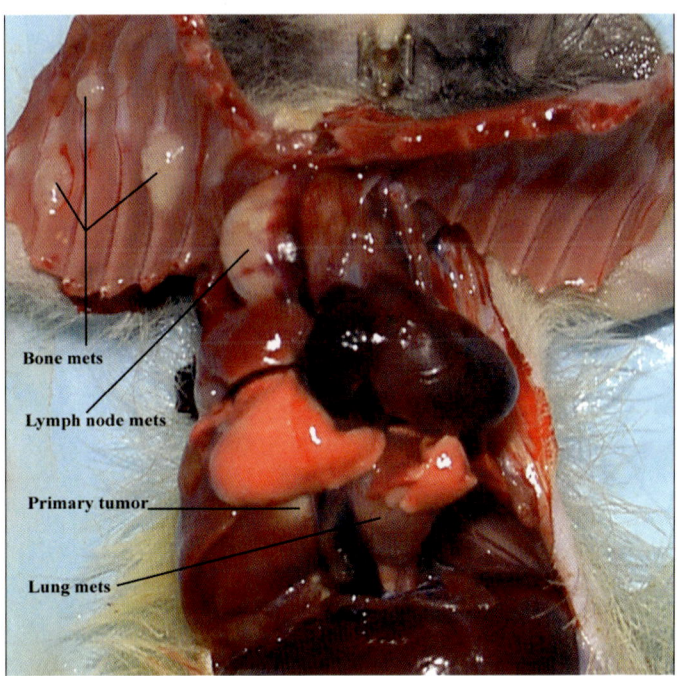

Figure 8. Thoracic cavity of a nude rat containing right caudal lobe tumor arising from NCI-H460 tumor fragments implanted endobronchially. Regional metastases to the mediastinal lymph nodes and systemic metastases to ribs and the left lung are shown.

Several other intrathoracic human lung cancer models have been described, all using immunocompromised mice. One is the traditional intravenous model in which the lung is colonized with tumor cells via the pulmonary circulation after tail vein injection (Kuo, Kubota et al. 1992; Kuo, Kubota et al. 1993). In the second, tumor grows in a sub-pleural location from fragments sewn onto the surface of the left lung through a thoracotomy (Wang, Fu et al. 1992). Recently, a SCID lung cancer model has also been described that develops lymphatic metastasis following percutaneous injection of cancer cells into the mouse lung (Miyoshi, Kondo et al. 2000). However, none of these models grow from a primary endobronchial site and none develop a consistent metastatic pattern in extrathoracic locations.

The NCI-H460 orthotopic rat model exhibits several advantages over other mouse models: (1) Primary tumors begin within the bronchial tree, where the majority of lung cancers normally originate. In the two mouse models the primary tumors originate either within alveolar capillaries or on the pleural surface of the lung. (2) In the rat model, there is no intentional injury to the lung. In the mouse thoracotomy model, a significant injury is associated with the chest wall incision and surgically suturing tumors to the lung surface; this likely causes release of various growth and angiogenic factors which may further modify a pleural environment already significantly different from that of a bronchus. (3) In the rat model, the primary tumors are confined to the right caudal lobe bronchus after implantation. This makes it very unlikely that any of the secondary tumor deposits arise from mechanical spread of the implanted tumor material, rather than true metastasis. In the mouse thoracotomy model, mechanical spread during implantation is a possible source of intrathoracic secondary tumor deposits. (4) In the rat model, the ten-fold larger animal size facilitates not only conventional surgical manipulations, such as cannulation, but is fundamental to the model, since the mouse bronchus is too small to accept the tumor fragments that appear to be necessary for expression of the metastatic phenotype.

In addition to using human cancer cell lines or their derived tumors for orthotopic implantation, histologically intact fresh, human lung tumor tissue or tissue from metastatic lesions can be orthotopically implanted. Such models putatively maintain intact critical stromal epithelial relationships, although the source of most, if not all, stromal tissue appears to be from the host rather than the original human xenograft (van Weerden and Romijn 2000). Very few such lung cancer models have been developed, in part because of technical obstacles and the generally poor growth of human lung cancer tissue in immunocompromised animals. Wang et al. (Wang, Fu et al. 1992) has developed such a model by surgical implantation of human small cell lung cancer tissue to mouse lung via a left thoracotomy. Metastases were found in contralateral lung and mediastinal lymph nodes. Progressive primary tumor growth and regional lymph node metastases are seen which closely resemble the original patient tumors histologically. Interestingly, one of these tumor lines developed contralateral lung metastasis in a fashion very similar to the patient from whom the tumor line originated.

Lung Cancer Models in Cancer Drug Development

Despite advances in basic cancer biology, animal models, especially human tumor xenografts, remain pivotal to cancer drug discovery and development. The value of a model depends on its validity, selectivity, predictability, reproducibility and cost (Zubrod 1972; DeVita and Schein 1973). Initially, lung tumor xenografts were designed with the intention of permitting patient specific chemotherapy. By learning the drug responsiveness of a particular xenograft, treatment of the patient from whom the transplanted material originated could be individualized. Unfortunately, variations in take-rate, the weeks to several months required for the transplants to grow, and the expense of maintaining xenografts make this strategy generally untenable in the clinical setting.

Early drug screening systems utilizing the L1210 or P388 mouse leukemia models represented a compound-oriented strategy. Any anticancer agent for clinical development had to prove itself in the murine leukemia/lymphoma models before further *in vivo* animal model development in a solid tumor panel. This resulted in a low yield of agents active against other solid tumor types. In order to develop screening systems with greater predictive power for the clinic, the U.S. National Cancer Institute (NCI) started to shift from a compound-oriented screen toward disease-oriented screens. NCI employs xenografts as an integral part of its drug discovery screening strategy (Khleif SN 1997). Drugs toxic to human cancer cell lines *in vitro* are tested on xenografts as a secondary screen. The *in vitro* studies permit high throughput screening and the cell lines found sensitive to a particular drug are used to choose appropriate xenografts for further testing. Lung tumor transplants often reflect the chemosensitivities of their tumors of origin. The growth of SCLC xenografts is inhibited by cisplatin, etoposide, cyclophosphamide, doxorubicin, and vincristine, while NSCLC grafts are much less responsive to those agents (Shoemaker RH 1988). Other animal tumor models can be selected to demonstrate a specific cytotoxic effect of the drug or biological agent. Primary lung tumors in mice can be used for screening effective single drugs and drug combinations prior to clinical testing. For example, cisplatin, administered by itself and in combination with indomethacin, decreases the size of NNK-induced carcinomas (Belinsky 1993). Although subcutaneous xenograft models such as the Lewis lung cancer system has been widely employed as an *in vivo* drug screen, the more complicated orthotopic models may be better suited for preclinical studies. Since orthotopic rodent tumors mimic biological aspects of clinical cancer (e.g. disease progression and metastasis) much better than do subcutaneous rodent tumors, orthotopic tumors are also likely to provide more relevant pharmacokinetic and pharmacodynamic information than subcutaneous tumors (Mulvin, Howard et al. 1993). Subcutaneous xenograft models have a long history in the pharmaceutical industry, and they are indisputably straightforward to use, however their record of accurately predicting the efficacy of anticancer agents in the clinic has been questionable.

A range of methods can be used to evaluate drug effects on tumors in animal models. Tumor size and tumor weight or volume changes are simple and easily reproducible parameters in subcutaneous xenograft models, but are more difficult, except at necropsy, in most orthotopic models. Morphologic changes and alterations in tumor immunogenicity or invasiveness are additional markers of response. Survival, perhaps the ultimate parameter, is a valid endpoint only if clinically relevant tumor progression is responsible for the animal's demise.

To accurately evaluate anticancer activity in an animal model system, validation of the model is critical. This entails the design of studies aimed at assessing tumor response to drugs or other agents known to have efficacy in patients with the particular type of cancer represented by the model. The H460 orthotopic lung cancer model has been validated (Johnston, Mullen et al. 2001) by treating tumor-bearing nude rats with one of four chemotherapy agents: doxorubicin, mitomycin, cisplatin, and the novel matrix metalloproteinase inhibitor, batimastat. The model shows consistent responses in the context of tumor weight, metastatic pattern and longevity to cisplatin and mitomycin treatment. The other two agents are largely ineffective, accurately reflecting the drug sensitivity patterns consistent with NSCLC and the H460 cell line. The model also detected cisplatin toxicity as assessed by body weight changes and kidney damage. A similar study was performed using two human lung cancers implantted in the pleural cavity of nude mice (Kraus-Berthier, Jan et al. 2000). Both studies show that selective cytotoxic agents may reduce primary tumor burden and prolong the survival of tumor-bearing animals. However, none of these agents are capable of completely eradicating tumor in these rodent models, reflecting the resistance of this disease to standard chemotherapy. In contrast, an orthotopic model of human small cell lung carcinoma (SCLC) demonstrates sensitivity to cisplatin and resistance to mitomycin C, reflecting the typical clinical situation (Kuo, Kubota et al. 1993). In contrast, the same tumor xenograft implanted subcutaneously responded to mitomycin and not to cisplatin, thus failing to match clinical behavior for SCLC. This suggests that the orthotopic site is crucial to a clinically relevant drug response. Similar phenomena have been observed, which underscores the potential effect of the microenvironment on drug sensitivity (Wilmanns, Fan et al. 1992).

A number of orthotopic nude mouse and nude rat models have been developed as *in vivo* preclinical screens for novel anticancer therapies that target invasion, metastasis and angiogenesis (Russell, Shon et al. 1991; Davies 1993; Furukawa 1993; Schuster 1993). A specific concern in studying anticancer agents with animal models derived from human cell lines is the degree of heterogeneity involved in the sample (Manzotti, Audisio et al. 1993; Price 1994). In other words, does serial passage of cell lines over months and years select out and propagate specific clonal elements of a tumor? Studies have shown that the molecular characteristics of both breast and lung cancer cell lines closely match their original human tumor (Gazdar, Kurvari et al. 1998; Wistuba, Bryant et al. 1999). From a phenotypic perspective, the H460 cell line does exhibit invasive and metastatic properties and maintains its drug sensitivity profile. However, other important characteristics, such as cytokine production or patterns of gene expression, may be lost or muted through serial passaging. Two potential solutions come to mind; either constructing model systems from fresh clinical tumor specimens and passaging the tumors serially as tumor lines, or creating multiple models representing all of the lung cancer histologies, thereby minimizing heterogeneity issues as much as possible.

Models for the Study of Lung Cancer Metastasis

The most remarkable feature of human lung cancer is tumor metastasis. It has been estimated that approximately 60% of cancer patients harbor overt or subclinical metastases at diagnosis, and it is the general consensus that the poor prognosis of lung cancer reflects the aggressive biologic nature of the disease. In particular, metastasis to mediastinal lymph nodes or distant organs produces poor prognosis in lung cancer. Unfortunately, very little is known about how lung cancer cells propagate distant metastasis and identification of molecules with a crucial role in the distant spread of lung cancer cells has been hampered by the absence of an appropriate experimental model system.

Intravasation and extravasation are two major steps for tumor cells to metastasize distantly. Entry of tumor cells into the circulation is the critical first step in the metastatic cascade, and although it has been assayed in various ways (Liotta, Kleinerman et al. 1974; Butler 1975; Glaves 1986), it has not been observed directly. Novel approaches that rely on the ability to specifically "mark" the tumor cell are promising. For example, one can engineer tumor cells to express the green fluorescence protein for *in vivo* fluorescence imaging. In order to understand the metastatic pattern of NSCLC, Yang M et al. developed a green fluorescent protein (GFP) expressing human lung cancer cell line H460-GFP. The GFP-expressing lung cancer was visualized to metastasize widely throughout the skeleton when implanted orthotopically in nude mice (Rashidi, Yang et al. 2000). This makes possible direct observation of tumor growth and metastasis as well as tumor angiogenesis and gene expression. This new assay is able to reveal the microscopic stages of tumor growth and metastatic seeding, superior to the previous transfection of lacZ to detect micrometastases (Lin, Pretlow et al. 1990; Boven 1992), as real-time visualization of micrometastases even down to the single-cell level becomes feasible.

In contrast to utilizing orthotopic implantation to enhance metastatic potential in lung cancer, the alternative approach of *in vivo* selection of metastatic tumor cell variants have also been applied. There is now wide acceptance that many malignant tumors contain heterogeneous subpopulations of cells with different potential for invasion and metastasis (Fidler and Hart 1982; Heppner 1984; Nicolson 1984; Nicolson 1987) and that metastasis results from the survival and proliferation of specialized subpopulations of cells that pre-exist within parental tumors (Fidler and Kripke 1977). The isolation of cell populations (from heterogeneous human tumors) that differ from the parent neoplasm in metastatic capacity provides a powerful tool with which to study those intrinsic properties that distinguish metastatic from nonmetastatic cells (Naito 1986; Morikawa 1988; Dinney, Fishbeck et al. 1995).

Efforts have recently been made to develop metastatic lung cancer cell variants through in vivo propagation and selection. New cell line variants, H460-LNM35 and H460SM were established through *in vivo* propagation of tumor cells derived from H460 tumor or lymph node metastases (Kozaki, Miyaishi et al. 2000; Blackhall, Pintilie et al. 2004). Selected variants of these tumor cells differ in their ability to metastasize compared to the parent cell line. This may provide a means of producing a highly metastatic orthotopic lung cancer model by direct cell implantation. Other opportunites involve the production of cell lines from transgenic models such as the K-ras hit and run allele, examining the potential of tumor engraftment in the absence of immune compromise. Selecting and enriching for metastatic variants constitute a useful model for the discovery and mechanistic evaluation of genes potentially involved in metastasis of human lung cancer.

Model organisms: new systems for modeling cancer

Although the mouse and rat have traditionally been used for *in vivo* modeling of cancer, a number of model systems are on the horizon that may impact the genetic dissection of tumor mechanisms and facilitate high-throughput screens for drug discovery. Model organisms such as the yeast *S. cerevisiae*, the nematode *C. elegans*, and the fruitfly *D. melanogaster* have been very productive in the genetic dissection of pathways in fundamental biologic and organogenic processes. Unfortunately none of these systems develops cancer. In contrast, the model organism zebrafish *danio rerio* does develop tumors with a variety of histologic subtypes that are similar to those present in humans. Fish have a long history of use in cancer toxicology studies because of this propensity to develop cancer. Because of considerable progress in zebrafish genetics and genomics over the past few years, the zebrafish system has provided many useful tools for studying basic biological processes. These tools include forward genetic screens, transgenic models, specific gene disruptions and small-molecule screens. By combining carcinogenesis assays, genetic analyses and small-molecule screening techniques, the zebrafish is emerging as a powerful system for identifying novel cancer genes and for cancer drug discovery (Stern and Zon 2003). Some of the advantages of zebrafish include ease and low cost of housing, large numbers of embryos produced from matings, ease of mutagenesis, and external, transparent embryos in which cleavage divisions, gastrulation, morphogenesis and organogenesis occur within 24 hours. Because of these advantages, the zebrafish has become a tour de force in vertebrate developmental genetics. Its' potential power and utility as a cancer model organisms is only beginning to be appreciated.

Summary

Many lung models are available, but unfortunately none accurately reflects all aspects of human disease observed clinically. Each has its own advantages and disadvantages that should be understood and evaluated prior to their use in addressing specific questions. In selecting the best model system, consideration should be given to the genetic stability and heterogeneity of transplanted cell lines, its immunogenicity within the host animal, and the appropriate biologic endpoints. There is increasing pressure on the research community to reduce, or even eliminate the use of animals in research. However, relevant animal model systems provide the appropriate interface between the laboratory bench and a patient's bedside for continued progress in cancer research and drug development. As in many other diseases, ever more sophisticated lung cancer models will be needed in the future as the complexities of this devastating disease are slowly unraveled.

Molecular Subtypes of Cancer from Gene Expression Profiling

Dennis A. Wigle and Igor Jurisica

Clinical and pathologic TNM staging

Clinical staging for cancer was first described by Pierre Denoix of France in the 1940's (Denoix 1944). The tumor-node-metastasis or TNM system he described was adopted by the International Union Against Cancer (UICC) in 1953 as the standard for cancer staging, serving as a common language for the description of cancer cases. Within this system, the "T" in TNM relates to tumour. It indicates tumour size, extent, or penetration (depth) of the tumour into surrounding normal tissue. The "N" stands for node, indicating the number of lymph nodes with cancer and/or the location of cancer-involved nodes. The "M" stands for distant metastasis, or spread of the cancer to other parts of the body, indicating cancer cells outside the local area of the tumour and its surrounding lymph nodes. The most common cancers using the TNM system are breast, colon and rectal, stomach, esophagus, pancreas, and lung. Other cancers staged with the TNM system include soft tissue sarcoma and melanoma. In total, staging systems exist for 52 sites or types of cancer. Some cancers are not staged using the TNM system, such as cancers of the blood, bone marrow, brain, and thymus.

In non-small cell lung cancer (NSCLC), the first description of a TNM classification system (Figure 9) was in 1974 by Clifford Mountain (MOUNTAIN, CARR et al. 1974). The description of clinical stages was based on a total of 1,712 NSCLC patients. An updated 1987 paper reviewed 3,753 cases with >2 year follow up to revise the system further (Mountain 1987). A 1997 revision was derived from ~5,000 cases, although it should be noted that little or no statistical data was used to back up many of the descriptors applied (Mountain 1997). The next revision is planned for 2009 by the International Association for the study of Lung Cancer (IASCLC), and will be based on a large cohort with an attempted statistical derivation for many of the T, N, and M descriptors incorporated.

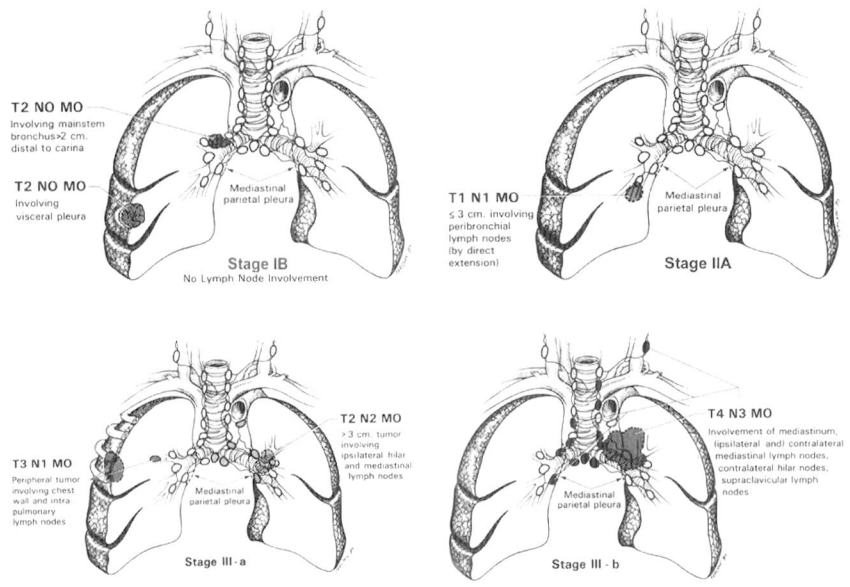

Figure 9. TNM staging system for NSCLC. Reprinted with permission from (Wigle, Keshavjee et al. 2005).

The importance of distinguishing clinical versus pathologic staging is also worth noting. The TNM system was originally designed to provide a rapid, simple method for assigning prognosis to a specific tumor based on the T, N, and M parameters. For many tumors, these parameters can be imprecise depending on the technologies applied to their determination. To use NSCLC as an example, T or N stage determined by plain chest x-ray is clearly different and far less accurate than determination by CT or PET-based imaging. Even these modern, sophisticated, imaging approaches are often inaccurate in comparison to the final stage determined at surgical resection via pathology. As a consequence, it is important to clearly make a distinction between clinical and pathologic staging. When dealing with clinical staging, information as to the modalities used to assign the stage are important to gauge the potential sensitivity and specificity of this testing. Molecular correlative studies require pathologic staging in order to ensure the highest degree of accuracy to the staging assigned. Importantly, the introduction of gene expression array technology enabled molecular staging of cancers, and its link to diagnosis and prognosis (Miyake, Adachi et al. 1999; McCann 2000; Wigle, Jurisica et al. 2002; Allgayer 2003; Marandola, Bonghi et al. 2004; Yardy, McGregor et al. 2005; Xi, Gooding et al. 2006). Figure 10 presents one example from NSCLC where samples with the same stage show different profile. Importantly, the profiles show significant correlation to recurrence and survival (Wigle, Jurisica et al. 2002). Similar trends can be detected when profiling each of the three NSCLC stage groups, as shown in Figure 11. The cross-stage molecular similarity of samples is more evident. In addition, patients in stage I with molecular profile similar to stage III patients died significantly earlier then expected for the group, and stage III patients similar to stage I profile survived about 3-4 times longer then expected for late stage NSCLC (Sultan, Wigle et al. 2002).

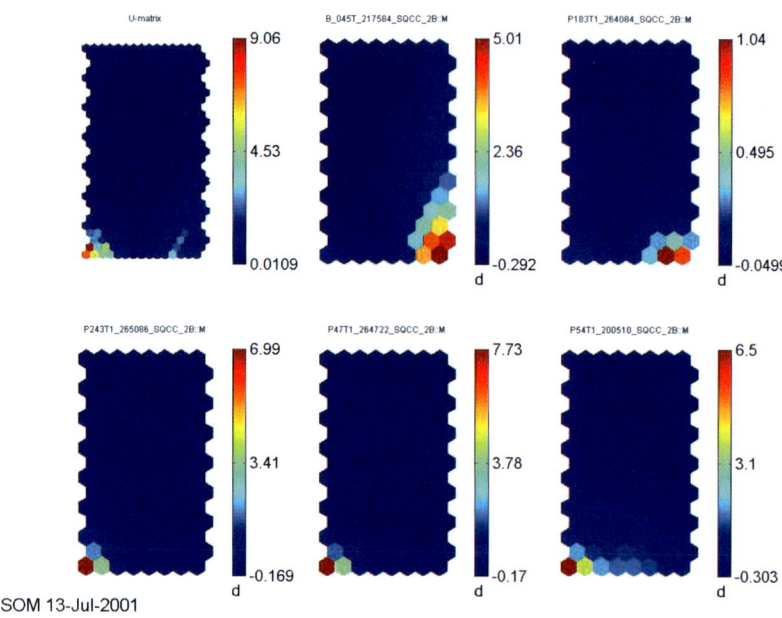

Figure 10. Molecular profiles for SQCC 2B samples from (Wigle, Jurisica et al. 2002), visualized using self-organizing maps (SOMs) (Kohonen 1995) in BTSVQ clustering program (Sultan, Wigle et al. 2002). The first map shows a generalized gene expression patters, mapped into a color scheme. Each other map shows representation of one sample, clearly the first two samples being different from the last three samples.

Molecular Subtypes of Cancer from Gene Expression Profiling   49

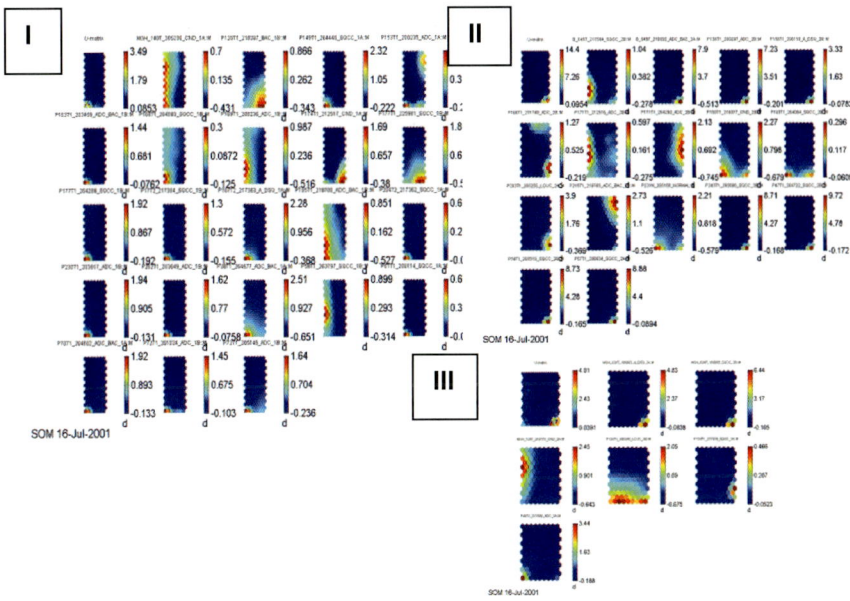

Figure 11. Molecular profiling of stage I, II, III groups of NSCLC samples from (Wigle, Jurisica et al. 2002), using self-organizing maps (SOMs) (Sultan, Wigle et al. 2002). The heat maps clearly show both pattern similar within the stage – but also across stages. Importantly, the across stages patterns correlate with survival.

In other tumor types, molecular adjuncts have been incorporated into the TNM staging system. In breast cancer, an I+ designation exists for pN0, signifying positive immunohistochemistry for cytokeratin markers in an otherwise normal lymph node based on histology. The mol+ designation also is used in the TNM classification for positive molecular findings based on cytokeratin RT-PCR in an otherwise histologically negative lymph node (Singletary, Allred et al. 2002). In prostate cancer, T1c cancers are those identified by biopsy performed because of an elevated PSA (Sobin LH 2002). No other solid tumor type has yet involved such molecular descriptors into the formal TNM staging system.

Gene expression profiling in cancer

The first reports of high-throughput gene expression profiling in cancer were published in 1999 with the description by Golub et al. regarding classification of acute myeloid leukemia (AML) and acute lymphoblastic leukemia (ALL) based on gene expression patterns alone (Golub, Slonim et al. 1999). Stemming from this, large studies were performed on most major tumor types using either oligonucleotide or cDNA-based microarrays. Oligonucleotide microarray technology popularized by Affymetrix has emerged over time to be a robust and reliable means to generate such data.

Early observations from many tumors demonstrated the potential for biologic classification of tumors into subgroups based on correlation with clinical outcome. This has been shown in preliminary data now from many tumor types, in some cases with molecular subtypes transcending traditional TNM stage classifications. The potential for molecular based staging to provide greater information than that available through current TNM systems has been a powerful driver for ongoing work in this area. Despite this promise however, clinically validated biomarker profiles are only now beginning to be tested in large patient cohorts to assess their translational utility. Using breast cancer as an example, gene-expression-profiling studies of primary breast tumors performed by different laboratories have resulted in the identification of a number of distinct prognostic profiles; however, many of these have little overlap in terms of gene identity. The earliest gene-expression profile test marketed in the United States is for early stage breast cancer. The Oncotype DX is a laboratory test that can be used on preserved (formalin-fixed, paraffin-embedded) stage I or II, estrogen receptor positive breast cancer tumor specimens from women whose tumors have not spread to their axillary nodes. Using the reverse transcription-polymerase chain reaction (RT-PCR), the test measures the level of expression of 21 specific genes to predict the probability of breast cancer recurrence. On the basis of those measurements, a "Recurrence Score" (RS) is assigned to the individual tumor. Measurements of five genes (Beta-actin, GAPDH, RPLPO, GUS, and TFRC) are used as controls. The other 16 genes include: genes associated with cell proliferation (Ki-67, STK15, Survivin, Cyclin B1, and MYBL2); genes associated with cellular invasion (Stromolysin 3, and CathepsinL2); genes associated with HER2 activity (GRB7 and HER2); genes associated with estrogen activity (ER, PR, Bc12, and SCUBE2); and three other genes with distinctly different activity in cancer cells (GSTM1, BAG1, and CD68). The RS is calculated by using a mathematical formula that includes the measured levels of these 16 genes to come up with a single RS between 1 and 100 for each individual tumor. The lower the score, the lower the predicted probability of disease recurrence (Paik, Shak et al. 2004). Although this test is available for molecular diagnostic testing, it has not been validated in a clinical trial format.

A recent report by Fan et al. (Fan, Oh et al. 2006) compared predictions from a single data set of 295 samples using five gene-expression-based models: intrinsic subtypes, 70-gene profile, wound response, recurrence score, and the two-gene ratio (for patients who had been treated with tamoxifen). Most of the models had high rates of concordance in their outcome predictions for the individual samples. In particular, almost all tumors identified as having an intrinsic subtype of basal-like, HER2-positive and estrogen-receptor-negative, or luminal B (associated with a poor prognosis) were also classified as having a poor 70-gene profile, activated wound response, and high recurrence score. The 70-gene and recurrence-score models, which are beginning to be used in the clinical setting, showed 77 to 81 percent agreement in outcome classification. The study concluded that even though different gene sets were used for prognostication in patients with breast cancer, four of the five tested showed significant agreement in the outcome predictions for individual patients and are probably tracking a common set of biologic phenotypes. This type of study remains to be repeated in other tumor types; however, it is a template for the kind of work that will be necessary to make clinical translation of expression signatures derived to date.

Lung cancer continues to be the most common cause of cancer-related mortality in both men and women in North America. It accounts for approximately 30% of all cancer deaths, a total greater than that from the next three cancers (breast, colon and prostate) combined (Society 2006). Despite this disease impact, the current staging system for lung cancer has remained largely unchanged for over 30 years, and continues to be based on histopathology and extent of disease at presentation (Mountain 1997). These classification systems alone have reached their limit in providing critical information that may influence treatment strategy. Clinical experience suggests that further subclassification and substaging of lung cancer remains possible given the heterogeneity of lung cancer patients at each disease stage with respect to outcome and treatment response. However, it is not currently understood if specific molecular subtypes exist within individual TNM stages, or if the TNM system is biased by the specific point in time at which a specific molecular subtype lesion is clinically identified. Answers to such questions are critical to our potential ability to incorporate molecular substaging into existing clinicopathological systems in a clinically relevant manner.

Recent studies have suggested that the application of microarray technology for gene expression profiling of NSCLC specimens may permit the identification of specific molecular subtypes of the disease with different clinical behaviour (Bhattacharjee, Richards et al. 2001; Garber, Troyanskaya et al. 2001; Beer, Kardia et al. 2002; Wigle, Jurisica et al. 2002; Bild, Yao et al. 2006; Potti, Mukherjee et al. 2006). Data from individual studies however, although large by microarray standards, have not been of the magnitude required to make significant inferences about the relationships between gene expression and clinical parameters. A recent study in non-small cell lung cancer has demonstrated the potential utility of gene expression information in the clinical management of lung cancer patients (Potti, Mukherjee et al. 2006). The current standard of treatment for patients with stage I NSCLC is surgical resection, despite the observation that nearly 30 to 35 percent will relapse after the initial surgery. The group of patients who relapse have a poor prognosis, indicating that a subgroup of these patients might benefit from adjuvant chemotherapy. In contrast, patients with clinical stage IB, IIA or IIB, or IIIA NSCLC currently receive adjuvant chemotherapy, but some may receive potentially toxic chemotherapy unnecessarily. Thus, the ability to identify subgroups of patients more accurately may improve health outcomes across the spectrum of disease. Although previous studies have described the development of gene-expression, protein, and messenger RNA profiles that are associated in some cases with the outcome of lung cancer, the extent to which these profiles can be used to refine the clinical prognosis and alter clinical treatment decisions is not clear. The Duke study identified gene-expression profiles that predicted the risk of recurrence in a cohort of 89 patients with early-stage NSCLC. They evaluated the predictor in two independent groups of 25 patients from the American College of Surgeons Oncology Group (ACOSOG) Z0030 study, a randomized trial of mediastinal lymph node sampling versus complete lymphadenectomy during pulmonary resection, and 84 patients from the Cancer and Leukemia Group B (CALGB) 9761 study, a prospective trial of tumor and lymph node collection during planned surgical resection in patients with clinical stage I NSCLC, with a primary objective to determine if occult micrometastases (OM) detected by immunohistochemistry (IHC) or real time PCR of CEA in histologically negative lymph nodes is associated with poorer survival. The gene expression model predicted recurrence for individual patients significantly better than did clinical prognostic factors and was consistent across all early stages of NSCLC. Applied to the cohorts from the ACOSOG Z0030 trial and the CALGB 9761 trial, the model had an overall predictive accuracy of 72 percent and 79 percent, respectively. The predictor also identified a subgroup of patients with stage IA disease who were at high risk for recurrence and who might be best treated by adjuvant chemotherapy. From this it was suggested that a randomized trial using gene expression profile as a randomization strategy for Stage IA patients may have utility in determining who might actually benefit from adjuvant therapy in early stage NSCLC.

The field anxiously awaits the results of a large, NCI Director's Challenge funded study looking at over 500 gene expression profiles of NSCLC using Affymetrix arrays, focusing on adenocarcinoma. This will be the largest microarray study yet in the public domain when published, and its' correlations of clinical outcome with gene expression biomarkers will provide a significant advance in information regarding biomarkers with the potential for clinical translation. It is hoped the study and its results will provide the volume of clinical data required to begin designing clinical trials incorporating gene expression biomarkers into clinical decision making.

Molecular subtypes within the TNM classification system

Many of the successes of modern cancer therapy, for example pediatric acute lymphoblastic leukemia, have been achieved using risk-adapted therapies tailored to a patient's risk of relapse. Critical to the success of this approach has been the accurate assignment of patients to different risk stratification groups. The heterogenous molecular alterations leading to ALL clearly have distinct responses to cytotoxic drugs, and this observation has been critical to tailoring therapies in a patient-specific manner (Golub, Slonim et al. 1999).

In non-small cell lung cancer, the current staging system has served as a simple and prognostically useful benchmark for the routine stratification of patients with this disease. It accurately predicts mean survival for patients based on tumor size and location, nodal status, and the presence or absence of systemic metastases. The existing system was originally derived from an analysis of each of these factors across a broad subset of patients to allow a common, systematic nomenclature for studying and guiding treatment decisions for lung cancer patients. Most of the advances occurring in lung cancer therapy have been derived on the foundation of this system. However, there remains significant heterogeneity in the outcome for any individual patient under this classification, a frequent observation amongst physicians responsible for treating these patients. We have speculated that similar to other tumor types, much of this patient-to-patient difference can be accounted for based on differing biology of individual tumors. As shown now from the major publications examining gene expression in lung cancer, significant heterogeneity can be observed both within and across clinical stages of this disease. From the observations described both from our data and from the datasets of Bhattacharjee et al., Garber et al., Beer et al., and Potti et al., it is clear that significant differences in gene expression do exist in lung cancers that do not correlate directly with clinical stage (Bhattacharjee, Richards et al. 2001; Beer, Kardia et al. 2002; Sultan, Wigle et al. 2002; Wigle, Jurisica et al. 2002; Potti, Mukherjee et al. 2006).This raises a number of questions about what the information from clinical staging is actually telling us about lung cancer biology. One possibility is that clinical stage is an accurate reflection of tumor biology, with less aggressive lesions found in early stages and later stages being comprised of more aggressive tumors. This is largely how NSCLC biology is perceived within clinical medicine. However, there are a number of lines of evidence against this idea. One, we know that significant heterogeneity in clinical outcome exists across large subsets of patients, which should not be the case if stage was truly reflective of biology. Two, expression profiling demonstrates that molecular heterogeneity exists both within and across clinicopathological stages. Three, the intuitive notion that one could detect a "bad biology" lesion early in the time course of disease progression, and hence label it a "good prognosis" tumor, compared with a "good biology" cancer detected at relatively late stages and hence labelled a "poor prognosis" tumor seems contradictory. The opposite end of the spectrum would be that clinical stage actually tells you nothing about biology at all. This cannot be true in the strictest sense given the relative accuracy of the staging system for predicting mean survival across large numbers of patients. In considering all the data available, we hypothesize that the current system is actually measuring the mean survival of a number of distinct molecular subtypes of lung cancer, each with its own inherent level of biology-based aggressiveness. The implications of this hypothesis would be that the staging system is actually measuring the point in disease progression that a lesion is discovered, and when this is averaged across all possible tumor biologies, it provides a relatively accurate picture of mean survival. However, for any one individual lesion with its own inherent and distinct biology, it is an inaccurate predictor of the likely outcome for that patient. The logical extension of this hypothesis is that further substaging of NSCLC patients based on tumor gene expression profile may be able to more accurately predict outcome and treatment response.

Recent studies involving gene expression profiling of clinical specimens have had a profound impact on cancer research. In many examples, correlations have been made between the expression levels of a gene or set of genes and clinically relevant subclassifications of specific tumor subtypes. These results have compounded expectations that true molecular classification and substaging of multiple tumor types may be possible, leading to measurable improvements in prognosis and patient management. Our initial observations based on statistical analysis and hierarchical clustering of gene expression data suggest a lack of correlation between the currently applied clinicopathological classification and staging system and gene expression profile. These findings are in contrast to the sharp delineation of groupings that has been demonstrated in other tumor types, such as malignant lymphoma (Alizadeh, Eisen et al. 2000). However, using a number of different analysis approaches, we are able to determine the presence of multiple molecular subtypes both within and across NSCLC clinical stages. Examination of expression studies using high density microarrays that have been performed to date in other solid tumor types appear to demonstrate a greater degree of heterogeneity overall in gene expression profiles compared to single or oligo-gene alteration tumors such as common leukemias or lymphoma subtypes. It is likely that a much greater number of expression profiles from clinical samples of solid tumors in general may be required to fully sample and delineate all existent molecular subtypes and begin to make the required epidemiologic correlations.

The results we describe suggest a number of implications for the classification and staging of non-small cell lung cancer. The absence of distinct gene expression profiles segregating with tumor stage implies significant heterogeneity in the biology of tumors both within and across discrete stages. However, as demonstrated by the relative accuracy of the current staging system to estimate mean survival, the point in time of disease progression at which a patient presents must have the largest overall effect in determining patient outcome across what may be multiple different biologic subgroups. Such biologic subgroups appear to be present as evidenced both by gene expression profile alone and in correlation with outcome data. In other words, stage as it is currently applied in NSCLC may actually reflect to some degree the effect of disease duration, and not represent a subtext of currently unmeasured molecular subtypes that are biologically and clinically relevant. The resulting implication is that a true molecular staging system, either built upon the current system or constructed anew, has the potential to further refine diagnosis, prognosis, and patient management for this lethal disease.

Where do we go from here?

One of the major issues with the current state of gene expression profiling in many fields is the lack of external validation for many of the biomarker panels that have been proposed. In many tumor types, high throughput gene expression profiling has lead to the derivation of a limited number of genes with correlation to clinical outcome, but not advanced beyond the model of using "training set" data applied to a "test set", frequently from the same group of experiments presented. Clearly not only external validation is required in patient volumes large enough to be believable, but also for samples collected in a prospective manner such that "clinical trial" type evidence exists justifying clinical translation of the results. No prospectively collected validation data incorporating a panel of biomarkers derived from microarray based gene expression studies currently exists in the literature. This level of evidence will clearly be required in all tumor types to move toward routine clinical use.

In NSCLC, a large volume of microarray data with associated patient clinical outcome data currently exists within the literature. This volume will be increased substantially with publication of the director's challenge results. Despite this however, no external validation studies have been performed to a degree to justify clinical applicability of any of the biomarker panels proposed. These vary in size from the "lung metagene" model of the Duke group, consisting of ~2,000 genes (Bild, Yao et al. 2006), to the 50 gene panel proposed by the Michigan group (Beer, Kardia et al. 2002), to our 3- and 6-gene predictors (Lau, Boutros et al. 2007; Zhu, Jurisica et al. 2007), and more recent Taiwan study with 5- and 16-gene predictors (Chen, Yu et al. 2007). It is possible that even smaller biomarker panels may have clinical utility as more studies emerge. Our analysis suggests that there are multiple minimal gene sets with comparable prognostic value (Lau, Boutros et al. 2007; Zhu, Jurisica et al. 2007). In any case, some form of prospective trial will have to be performed to facilitate true, evidence-based, clinical translation of the results. Although this may seem a long way off at present, it really represents only a starting point for the molecular staging of NSCLC. An even greater volume of work will be required to tell us how best to use these biomarker panels in clinical practice. Potential applications in NSCLC are most obvious at the extremes of the current staging system. This includes decisions regarding chemotherapy in early stage disease, and the suitability of trimodality approaches in stage IIIA disease. Refinements beyond these two important questions await further study.

Mass Spectrometry-based Systems Biology

Thomas Kislinger

Disciplines of Systems Biology

Systems biology is a new biological discipline aiming at the global detection of gene products and metabolites in a qualitative and quantitative manner. The three main branches of systems biology are:

1. **Transcriptomics**: The large-scale detection of mRNAs in biological samples. Microarray or gene chip technology is utilized to globally measure changes of the transcriptome under various biological conditions.
2. **Proteomics**: The large-scale detection of proteins in a biological sample.
3. **Metabolomics**: The systematic detection of small molecular metabolites in a biological organism.

We will focus our attention to proteome research and further describe the sub-branches of proteomics currently investigated.

Sub-specialties of proteomics

There are four main specialties in proteomics:

1. **Expression proteomics**: The ultimate goal of expression proteomics is the generation of global "snap-shots" of protein expression patterns in any given biological samples in a qualitative and quantitative manner.

2. **Functional proteomics**: Aims at the large-scale detection of protein-protein interactions in any given organism.
3. **Proteomics of posttranslational modifications**: This branch of proteomics intends to detect every posttranslational modification (PTM) present in a biological sample. Equally important is the accurate quantification and exact localization of PTMs (e.g. which amino acid carries the modification).
4. **Structural proteomics**: The ultimate goal of solving the three dimensional structure of every known protein.

In the long run the integration of results from all sub-specialties of proteome research will enable researchers to obtain "detailed pictures" of the physiological conditions of cells, tissues or organisms of interest. The systematic monitoring of changes in these "fingerprints" of protein expression under various biological conditions, e.g. development, stress, and disease will allow biologists to better understand fundamental biological processes.

Mass spectrometry-based proteomics

Historically the term "proteome" was first introduced by Wasinger et al. in 1995 to describe the complete set of proteins expressed by a given organism (Wasinger, Cordwell et al. 1995). Modern proteomics can be considered as genome-wide protein biochemistry with the aim of studying and detecting all proteins in a biological system at the same time. Due to the tremendous potential, this new discipline of biology (a.k.a. "omics") has generated an enormous hype within the biological research community (Domon and Aebersold 2006).

The central workhorse of proteome research is the mass spectrometer (MS). Mass spectrometry is the analytical technique used to measure the mass-to-charge ratio of ions in the gas phase. A MS in general consists of three parts: the ion source, the mass analyzer and the ion detector. The ion source is where the analyte is ionized. There are a multitude of different ionization techniques, but in biological mass spectrometry electrospray ionization (ESI) and matrix-assisted laser desorption/ionization (MALDI) are the most commonly used ionization techniques (Karas and Hillenkamp 1988; Fenn, Mann et al. 1989). The development of these two ionization techniques in the late 1980s was a major breakthrough for biological mass spectrometry and was awarded the 2002 Nobel Prize in Chemistry.

After ions are formed and transferred into the gas phase, their mass-to-charge ratios are measured by the mass analyzer. Basically there a five different mass analyzer currently in us. This includes the time-of-flight (TOF), the quadrupole, the ion trap, the Fourier transform ion cyclotron resonance (FT-ICR) and the Orbitrap mass analyzers. All mass analyzers use either magnetic or electric fields to separate the generated ions according to the mass and charge (m/z). The final component of a MS is the detector. A detector in general records the signal that is generated when an ion hits its surface. Typically MS detectors are some kind of electron multipliers, amplifying the signal generated by each ion as it hits the detector.

Electrospray ionization - ESI

ESI was initially developed by John Fenn and coworkers (Fenn, Mann et al. 1989). It is the most frequently used ionization technique for large biomolecules, because of its mild ionization properties. This prevents the usual fragmentation of large molecules when ionized. The molecule to study is pushed though a very small glass or metal capillary. The liquid contains the analyte of interest as either positively or negatively charged ions. A strong electric field is applied to the buffer solution. At the tip of the capillary a fine aerosol of small droplets is formed. The analyte of interest is dissolved in these droplets and as the solvent begins to evaporate, the charged molecular ions will be forced closer together. Eventually, the repelling force between these similarly charged ions becomes so strong that the droplets explode, leaving behind analyte ions free of solvent. These ions will enter the MS and their m/z ratio will be measured in the mass analyzer.

ESI is the primary ionization form used in liquid chromatography mass spectrometry (LC-MS). In proteomics, complex peptide mixtures are first separated by LC and eluting peptides are directly ionized into the MS (termed: shot-gun proteomics). The MS records the m/z for every eluting peptide. To identify peptides present in a biological sample, a second dimension mass spectrum, the so call tandem mass spectrum (MS/MS) is required. Briefly, individual peptide ions are collided with inert gas molecules (helium) resulting in the generation of sequence specific peptide fragmentation patterns, in a process called collision induced dissociation (CID). The breakthrough in ESI-LC-MS came with the introduction of nano-electrospray LC-MS. Nano-electrospray is carried out in narrow fused silica capillaries (inner diameter 50-150 μm) at flow rate in the range of several hundred nanoliters per minute (200-400nl/min). This drastically improved the detection range of proteins in complex biological samples and enabled modern proteomics. The most commonly used mass analyzers of modern proteomics laboratories are the triple quadrupole and the ion-trap. Especially the ion-trap mass analyzer is considered the "work horse" of proteomics, due to their robustness, low maintenance and relatively low initial price. The recent commercial introduction of the linear ion-trap mass analyzer further improved the utility of these popular mass spectrometers. The major advantage of the linear ion-trap is its greater ion trapping efficiency and faster ion ejection rates. This results in a greater sensitivity and larger number of recorded spectra. The end result is a larger number of identified proteins in complex biological samples. One of the negative points of ion-traps is their relatively low mass accuracy and resolution. In this perspective, Fourier transform ion cyclotron resonance (FT-ICR) and Orbitrap mass analyzers provide the best performance. Both mass analyzers have mass accuracies in the low ppm range (10 ppm and better) and resolutions greater than 50,000. This can greatly enhance protein identifications, although at the cost of much higher purchasing costs. Most recently, Thermo Fisher Scientific (formerly Thermo Finnigan) has combined linear ion-trap mass analyzers with FT-ICR and/or Orbitrap mass analyzers in hybrid mass spectrometers. These instruments possess the advantages of both worlds; high scanning speed and sensitivity of the linear ion-trap and high mass accuracy and resolution of the FT-ICR and Orbitrap mass analyzer (Aebersold and Mann 2003; Steen and Mann 2004; Ong and Mann 2005).

Multidimensional protein identification technology – MudPIT

MudPIT was initially developed by John Yates 3rd and co-workers. It is basically an on-line nano-electrospray two-dimensional microcapillary chromatography coupled directly to a MS (Figure 12) (Link, Eng et al. 1999; Washburn, Wolters et al. 2001; Wolters, Washburn et al. 2001). Briefly, microcapillary chromatography columns, as described above, are packed with two orthogonal chromatography resins. The first dimension consists of a strong cation exchange resin (SCX) and the second dimension is a reverse phase 18 (RP-18) resin. Complex protein mixtures (e.g. whole cell extract) are enzymatically digested using sequence specific enzymes (e.g. trypsin, endoproteinase Lys-C) to generate very complex peptide mixtures. These are loaded directly onto the biphasic microcapillary column, using a pressure vessel. Under acidic conditions peptides will preferentially bind to the SCX resin, which will serves as a peptide reservoir. Peptide separation is accomplished by running multi-step, multi-hour separation sequences. Briefly, each sequence consists of multiple independent steps. Each step starts with a "salt bump" of ammonium acetate pushing a subset of the peptides bound to the SCX onto the RP-18 resin. Peptides on the RP-18 resin will be separated by water/acetonitrile gradients and directly ionized into the MS. In the following step of a MudPIT sequence the salt concentration of the ammonium acetate "bump" will be increased in concentration, moving the next set of peptides onto the RP-18 resin. This will be repeated until the reservoir of peptides bound to the SCX resin is completely depleted. The MudPIT technique is a very powerful tool for expression proteomics of complex biological samples and allows for much deeper proteomics detection depth (= number of identified proteins). The utility and usefulness of MudPIT has been demonstrated in recent years by the publication of several high impact papers. Several key papers are highlighted below.

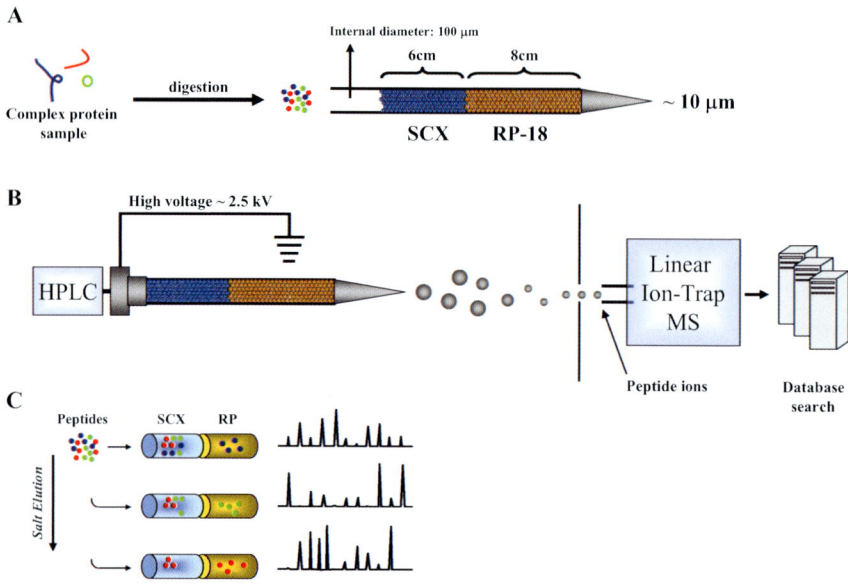

Figure 12. Multidimensional protein identification technology (MudPIT). (A) Complex protein mixtures are digested to peptides which are loaded onto microcapilary columns containing two chromatography resins. (B) Columns are place in-line with a HPLC pump and directly eluted into the mass spectrometer. Generated spectra are searched on computer clusters. (C) Shown is the basic concept of multi-step MudPIT runs. In each step a "salt bump" is used to move a sub-set of peptide bound to the SCX onto the RP resin. These peptides are then chromatographically separated and directly eluted into the MS. In the next step the salt concentration is increased to move another set of bound peptides from the SCX resin onto the RP resin.

The original publication by Washburn and Wolters applied MudPIT to the analysis of complex yeast extracts. The authors reported the confident identification of 1,484 proteins, a number much higher than anything reported by two-dimensional gel electrophoresis (2-DE) (Washburn, Wolters et al. 2001; Wolters, Washburn et al. 2001). Especially impressive was the detection of low abundance proteins, such as sequence specific transcription factors and protein kinases. Additionally, MudPIT was able to identify 131 transmenbrane proteins with three or more transmembrane domains, a protein class notoriously difficult to identify by gel based proteomics technologies. Since this landmark publication in early 2001, the group of John Yates 3rd and several other investigators have applied and improved the powerful MudPIT technique. Koller et al. presented a systematic proteomics analysis of three different tissues (leaf, root and seed) of the commercially important grain Oryza sativa (rice) (Koller, Washburn et al. 2002). By combing 2-DE and MudPIT a total of 2,528 proteins were detected, many of which in a tissue-specific manner. Several known allergenic proteins were detected specifically in the rice seed, demonstrating the potential of proteomics technologies in the monitoring of food products. In 2002, Florens et al. published the first proteomics investigation of the life cycle of Plasmodium falciparum, the malaria parasite (Florens, Washburn et al. 2002). The utility of expression proteomics in genome annotation was impressively demonstrated by the detection of over 1,200 hypothetical proteins, previously described as open reading frames. In 2003, Wu and others described a comprehensive MudPIT-based analysis strategy for membrane proteins (Wu, MacCoss et al. 2003). By applying several smart biochemical preparations, the authors were able to facilitate the identification of membrane proteins, putative posttranslational modifications and the characterization of membrane protein topology. This analysis strategy was applied to investigate the proteome of the stacked Golgi fraction (Wu, MacCoss et al. 2004). The study identified 41 proteins of unknown function and identified arginine dimethylation as a post-translational modification of Golgi proteins. Schirmer et al. reported the identification of nuclear envelope proteins with potential implication in dystrophies (Schirmer, Florens et al. 2003). The authors applied "subtractive proteomics" as a smart trick to distinguish bona fine nuclear proteins form potential cross contaminating proteins. Briefly, the authors used ultracentrifugation in density gradients to isolate nuclear envelopes. In parallel microsomal membranes were isolated and both compartments were extensively analyzed by MudPIT. Proteins uniquely found in the nuclear envelope (NE) isolation were shown to be highly enriched in true, known NE proteins. Importantly, this subset of proteins contained 67 uncharacterized open reading frames with predicted transmembrane domains. 23 of these hypothetical NE proteins mapped into chromosomal regions previously linked to a variety of dystrophies. In 2003, Kislinger et al. published one of the first MudPIT-based proteomics investigations of mammalian tissues (Kislinger, Rahman et al. 2003). The authors analyzed several specifically isolated organelle fractions (cytosol, membranes, mitochondria, nuclei) from mouse liver and lung. A detailed analysis strategy, termed PRISM (Proteomic Investigation Strategy for Mammals) was developed. PRISM includes subcellular fractionation of mammalian tissue, extensive MudPIT-based proteome profiling, statistical validation of generated search results to minimize the false discovery rate and automatic mapping to available Gene Ontology terms, to streamline the datamining process. The authors reported the confident identification of over 2,100 proteins in a tissue and organelle specific manner. Just recently, the same group of researchers reported an extension of this work by comprehensively analyzing organelle fractions from six healthy mouse tissues (brain, heart, kidney, liver, lung and placenta) (Kislinger, Cox et al. 2006). Almost 5,000 proteins were confidently detected. By applying sophisticated bioinformatics and machine-learning algorithms the subcellular localization of over 3,200 proteins could be determined with high confidence.

Schnitzer and colleagues have recently published a very interesting biological application of the MudPIT technology. The authors applied the silica-bead perfusion technique to selectively investigate the luminal plasma membrane proteome of rat lung microvasculature endothelial cells (Durr, Yu et al. 2004). Over 450 proteins, highly enriched in known plasma membrane components, could be identified. Interestingly, by comparing the *in vivo* identified proteins with endothelial cells cultured *in vitro*, large differences were detected, arguing that tissue microenvironment is a regulating factor of protein expression patterns.

More recently several technical considerations have been described in the scientific literature. Saturation of detection and random sampling are two important issues to consider when dealing with MudPIT-based profiling of very complex protein mixtures. Random sampling describes the incomplete acquisition of peptide spectra in very complex samples. In complex mixtures the MS is not capable of recording mass spectra for every eluting peptide (Liu, Sadygov et al. 2004). Repeat analysis of the same sample is highly recommended to achieve a certain level of saturation, and statistical models have been developed to help predict the number of repeat analysis required to achieve a certain level of saturation.

Matrix-assisted laser desorption/ionization - MALDI

MALDI is a "soft" ionization technique used to study large biomolecules by MS. The methodology was initially introduced by Karas and Hillenkamp in 1985 (Karas and Hillenkamp 1988), as a further improvement of the original procedure described by Koichi Tanaka (Tanaka, Waki et al. 1988). Briefly, an analyte of interest is mixed in large excess with matrix molecules and spotted onto a stainless steel target. A pulsing nitrogen laser is fired at the analyte-matrix co-crystal, resulting in ionization of both molecules. The matrix, usually an aromatic acid, is used to protect the analyte molecule from destruction by the laser. Commonly used matrices are, sinapinic acid, α-cyano-4-hydroxycinnamic acid and 2,5-dihydroxybenzoic acid. A high electric field is used to accelerate generated molecular ions into the MS. In general, time-of-flight (TOF) mass analyzers are used to determine the m/z of each molecular ion, hence the term MALDI-TOF-MS.

Traditionally, MALDI-TOF-MS was coupled to electrophoretic separation of proteins. Protein mixtures were first separated by one- or two-dimensional gel electrophoresis, depending on the complexity and the desired resolution. Gels are stained with either silver or coomassie to visualize the separated proteins. Spots of interest are excised, in-gel digested and analyzed by MALDI-TOF-MS. By using sequence specific proteases (e.g. trypsin which cleaves proteins C-terminal to the amino acids lysine and arginine) each unique protein will generate a different set of peptides that can be used for identification. This technology called "peptide mass fingerprint" (PMF) and has been successfully applied for the identification of proteins by MALDI-TOF-MS. Drawbacks of gel-based MALDI-TOF-MS peptide fingerprinting are that comprehensive proteome coverage is rarely achieved, it is very time consuming and unambiguous protein identification of proteins from higher mammalian species cannot always be achieved based on a peptide fingerprint alone. Recent improvements in MALDI-TOF technologies, especially the introduction of the TOF/TOF mass analyzer, capable of recording tandem mass spectra have overcome some of these limitations. Recently, MALDI-TOF-MS has been coupled to separations by nano-LC. Briefly, complex peptide mixtures are separated by nano-LC and directly eluted onto discrete spots on a sample target plate. Each spot on the target corresponds to a defined chromatographic retention time. After addition of matrix solution each spot is analyzed by MALDI-TOF-MS. A nice feature of LC-MALDI-TOF-MS is that target plates can be stored for re-analysis at a later time.

Protein identification in proteome research

The identification of proteins in biological samples from mass spectral data is a central task of proteome research. Traditionally, skilled biologist's manually interpreted individually recorded mass spectra, in a process called *de novo* sequencing (Dancik, Addona et al. 1999; Gevaert and Vandekerckhove 2000). Obviously, the results were highly dependent on the skill of the interpreter and the quality of the mass spectrum. Modern proteomics projects generate 1000s to 100000s of mass spectra, clearly limiting the success of de novo sequencing efforts. Today a multitude of public and commercial search algorithms are available to the proteomics research community, significantly speeding up the process of protein identifications. In the next couple of paragraphs will summarize the most commonly used MS search algorithms. We also encourage the reader of this article to consults book chapters and reviews specifically dealing with database search algorithms (Figure 13) (Steen and Mann 2004).

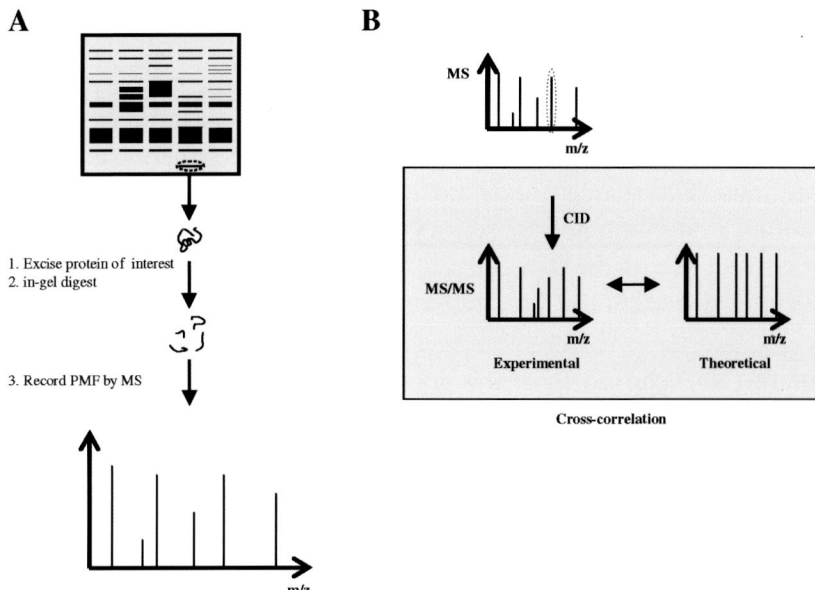

Figure 13. Protein identification by mass spectrometry. (A) Proteins are separated by one-dimensional gel electrophoresis and bands of

Identification of proteins from peptide mass fingerprints

Peptide mass fingerprinting (PMF) is a technique that matches peptide masses generated by the enzymatic digest of proteins to their theoretical masses generated from a protein sequence database. In the first step an unknown protein (e.g. excised from a gel or biochemically purified) is digested with a sequence specific enzyme (e.g. trypsin, Lys-C, Glu-C) to generate peptides. The basic idea of PMF is that every unique protein will generate a unique set of peptides and hence peptide masses. This information is used by search algorithms to identify the unknown protein in the sequence database. Commonly used algorithms are:

- Aldente (http://ca.expasy.org/tools/aldente/)
- ProteinProspector (http://prospector.ucsf.edu/)
- PROEL (http://prowl.rockefeller.edu/)
- Mascot (http://www.matrixscience.com/)

Basically, a list of identified peptide ion masses is uploaded into the search algorithm. The biologist then specifies a set of user defined parameters such as, enzyme used to generate the PMF, potential modifications, mass of the analyzed proteins, protein sequence database used for the search and mass accuracy of the measured peptide ions. The search algorithm will return putative protein identifications along with an algorithm specific score describing the quality or confidence of the identification. Obviously, several caveats could potentially complicate protein identification by PMF. The mass accu-

◄─────────────────────────

interest are excised from the gel and in-gel digested. The generated peptides are analyzed by MALDI-TOF-MS to generate a peptide mass fingerprint (PMF). (B) Protein identification by tandem mass spectrometry. First, the m/z of parent ions is recorded. Then individual peptide ions are isolated and fragmented by collision induced dissociation. Cross-correlation of theoretical MS/MS spectra generated by the search algorithm based on the mass of the parent ion with the experimental tandem mass spectra is used to identify the peptide sequence.

racy of the peptide ion masses clearly influences the identification process, as does the complexity of the analyzed protein samples. A mixture of several proteins rarely leads to a confident identification by PMF. Unknown protein modifications or absence of the protein sequence in the database can also cause difficulties for protein identification.

As a general rule, PMF works well for the routine identification of 2DE separated proteins from less complex organisms such as E. coli or yeast. Proteins from complex mammalian organisms (mouse, human) are better analyzed by tandem mass spectrometry (see below).

Protein identification by tandem mass spectrometry

The sequencing of peptides was an enormously painful process, usually done by Edman sequencing, before the development of tandem mass spectrometry (Edman 1960; Edman and Begg 1967). As described above, tandem mass spectrometry is the selective isolation of a precursor or parent ion, followed by collision induced dissociation. The generated MS/MS spectrum is specific to the amino acid sequence of the peptide and can therefore be used for its identification. The main barrier to the wide spread use of tandem mass spectrometry for the identification of proteins was the difficulty of spectral interpretation. Especially modern proteomics laboratories, generating 1000s of MS/MS spectra per day, are heavily dependent on automatic spectral identification by database search algorithms. There are several well established commercial and open source algorithms available to the proteomics researcher.

The SEQUEST algorithm, originally developed by the group of Dr. Yates 3rd and now commercially available through Thermo Fisher Scientific is the most commonly used code (Eng, McCormack et al. 1994). Briefly, candidate peptide sequences are pulled from the protein sequence database of choice, based on the m/z of the parent ion. Theoretical tandem mass spectra are generated for each of these peptides and cross-correlation is used to compare theoretical spectra to the experimental spectrum. The best matching peptide is reported to the biologist.

Several other algorithms are in wide use. These include Mascot, commercial software from Matrix Science (http://www.matrixscience.com/). It is a variation of the original Mowse code, developed by Pappin and Perkins (Perkins, Pappin et al. 1999). More recently, open source algorithms have been made available to proteomics researchers. These include X!Tandem from the Global Proteome Machine Organization (http://www.thegpm.org/TANDEM/index.html) (Craig and Beavis 2004) and OMSSA from the NCBI (http://pubchem.ncbi.nlm.nih.gov/omssa/) (Geer, Markey et al. 2004). Both algorithms are excellent alternatives to the very pricy SEQUEST and Mascot algorithms, and run on most commonly used operating systems (Windows, Mac OS X and Linux).

Filtration of search results

Comprehensive proteome projects using MudPIT-type profiling on linear ion-trap mass spectrometers, are capable of recording several 100,000s mass spectra in a single day. After searching this data with the described database search algorithms several hundred proteins are identified. A pressing issue of proteomics is the filtration of these search results to ensure high quality data. Basically, the art of proteomics is not to generate long lists, but to report comprehensive proteome profiles of high confidence and integrity. The ultimate goal is to rigorously minimize false positive identifications without generating too many false negatives. In other words, we are searching for stringent, objective filter criteria without throwing the baby out with the bathwater. In recent years several statistical algorithms have been developed allowing for an objective filtration of generated search results (Keller, Nesvizhskii et al. 2002; MacCoss, Wu et al. 2002; Kislinger, Rahman et al. 2003; Nesvizhskii, Keller et al. 2003; Peng, Elias et al. 2003). A statistical confidence of correct peptide/protein identification is reported to the user, which allows filtering the data to obtain an acceptable false discovery rate.

Another trick frequently applied in proteome research is to search the MS-data against protein sequence databases containing an equal number of "dummy decoy sequences". These decoy sequences are generated by inverting the amino acid sequence of every protein in the native target database. This will generate protein sequences unlikely to exist and search result returning these proteins are considered false positive identifications. Filter criteria are applied to minimize the number of decoy sequences (Kislinger, Rahman et al. 2003; Peng, Elias et al. 2003). Another recently applied trick to minimize false positive identifications is the use of high mass accuracy MS. First the mass of a peptide is measured with high accuracy by MS, using a FT-ICR-MS or an Orbitrap. Second the theoretical mass of this peptide identified by the search algorithm is calculated. The mass difference should be no larger than 20 ppm for correct peptide identifications (Haas, Faherty et al. 2006; Yates, Cociorva et al. 2006).

Protein quantification in proteome research

In proteomics we are not just interested what proteins are present in a sample, but how much of a protein is there and how does its abundance change under certain conditions (e.g. development, disease, after treatment etc.).

Protein quantification is a very challenging task, especially for complex mixtures, and most definitely not every protein will be quantified. Several methodologies have been developed in recent years and we will describe the most important technologies below.

Protein quantification with stable isotopes

Relative protein quantification by stable isotopes is the most commonly used methodology. Briefly, one sample is labeled with a heavy isotope (e.g. C13 or N15) and the other sample is labeled with the corresponding light isotope (C12 or N14). The two samples are mixed and analyzed by MS. In LC-MS the two differentially labeled proteins will have the same retention time and co-elute together into the MS. The MS will separate the two species based on their different mass. By comparing the elution profile of the two peptides over time and integrating the area under the curve of both peptides relative quantification is achieved. Several different isotope labels are commercially available. These include the ICAT (isotope coded affinity tag) (Gygi, Rist et al. 1999) and the ICPL (isotope coded protein label) (Schmidt, Kellermann et al. 2005). Briefly, these labels chemically modify a specific functional group on the side chain of an amino acid. The ICAT label specifically reacts with the thiol-group of cysteines and the ICPL label specifically modifies the epsilon-amino groups of lysines.

Other common forms of isotope labeling are based on metabolic labeling of cells. The cell growth medium is supplemented with amino acids or other essential nutrients in either the heavy or light isotopic form. Cells grown in this medium will either incorporate the heavy or light isotopes into their proteins. Mann and co-workers developed this form of isotope labeling and termed it SILAC (stable isotope labeling in cell culture) (Ong, Blagoev et al. 2002).

Label-free protein quantification

Basically there are two label-free methods for protein quantification in shot-gun proteomics. The advantage of label-free quantifications is that no expensive stable isotope reagents are required. This is a significant cost advantage as isotope labels such as ICAT or ICPL are extremely expensive.

1. **Peak integration:** In this method two samples are resolved by two separate LC-MS runs and appropriate peaks are quantified by integrating the area under the peak in each of the two runs. The success of this methodology is highly dependent on the reproducible separation of peptide mixtures by nano-flow chromatography (reproducible retention time of peptides). Furthermore, the mass resolution of the MS is highly important (to make sure the same peaks are integrated) (Callister, Barry et al. 2006; Ono, Shitashige et al. 2006; Wang, Wu et al. 2006).

2. **Spectral counting method:** More recently several investigators have independently demonstrated that spectral counting (the number of high quality spectra recorded for a given peptide)

accurately reflects protein abundance (Liu, Sadygov et al. 2004; Fang, Elias et al. 2006; Kislinger, Cox et al. 2006). This method is very accurate if big differences in relative protein abundance are measured and less accurate for small changes. Furthermore, by comparing spectral counting to metabolic labeling of proteins using stable isotopes, spectral sampling was proven to be more reproducible and covering a wider dynamic range.

Gel based protein quantification

Still a widely used method for the relative quantification of proteins is the selective staining of proteins after electrophoretic separation. This strategy has been widely used in combination with two-dimensional gel electrophoresis (2-DE) (Gorg, Obermaier et al. 2000). Briefly, protein mixtures are first separated by 2-DE, separating proteins based on the isoelectric point in the first dimension and based on the molecular mass in the second dimension. Gels are then stained to visualize protein separation. Several different stains are available, ranging from silver staining and coomassie staining to diverse fluorescence stains. Sophisticated computer tools exist that scan, compare and quantify individual spots based on the intensity of the stain. DIGE (differential in-gel electrophoresis) is a further improvement of this principle. Two protein samples are first labeled with two fluorescence labels (Cy3 and Cy5), mixed and separated by 2-DE. Quantification is rapidly and accurately achieved based on the fluorescence intensity of the individual label (Yan, Devenish et al. 2002).

Irrespectable of what method is used for relative protein quantification we recommend caution in the interpretation of the results and validation by independent methods (e.g. Western blotting).

Application of proteomics to cancer research

Despite major efforts and financial investments in cancer research, both on the clinical and basic research side, cancer remains a major health risk. According to the Canadian Cancer Society (Canadian

Cancer Statistics 2006) an estimated 153,100 new cases of cancer and 70,400 deaths from cancer will occur in Canada in 2006. Importantly, inadequate measures for early detection of this devastating disease exist, which could help to cure or prevent the disease from further progression. Systems biology and especially proteomics could help to gain new insight into cancer biology and/or identify diagnostic biomarkers of cancer.

In the following sections we will review some of the major developments and applications of proteomics in cancer research. For further reading we encourage the readers to consult some of the excellent reviews published in cancer biology in recent years.

SELDI-TOF-MS-based cancer profiling

Surface-Enhanced Laser Desorption/Ionization Time-of-Flight mass spectrometry is basically a variation of MALDI-TOF-MS. The system was commercially introduced by Ciphergen Biosystems. In principle it is a low resolution time-of-flight mass spectrometer that uses sample target plates coated with a variety of chromatography resins (e.g. ion exchange, reverse phase, metal ion binding etc.). Briefly, complex biological samples (serum, tissue lysates) are directly applied to the Cipergen ProteinChip, coated with an ion chemistry of choice. After incubation a selective wash step removes unbound material and crucially reduces the sample complexity, as only analytes interacting with the resin of choice will be retained on the chip and analyzed by TOF-MS. A simple mass spectrum containing the m/z and intensity values of proteins and peptides present in the sample is recorded. The results can be viewed by the user applying several software suites developed by Ciphergen. Sophisticated patter recognition algorithms are used to detect significant differences in protein patterns to distinguish between samples (e.g. healthy vs. disease). Unfortunately, the low mass resolution achieved and inability to perform tandem mass spectrometry by SELDI-TOF-MS is insufficient in identifying the identity of a particular protein. More recently some of these initial limitations were overcome by coupling SELDI ProteinChips to high resolution MS (QStar from Applied Biosystems) including the capability of recording MS/MS spectra. One of the big advantages of SELDI-TOF-MS, especially as a clinical diagnostic tool, is its high sample throughput. Sample binding, washing and MS analysis are highly automated, by smart robotic systems. With the continuous improvements in MS technologies and data analysis/mining tools we believe that SELDI-TOF-MS will continue to be a useful tool, especially in clinical diagnostics. For a very detailed review on the SELDI-TOF-MS technique we highly recommend the review by John Roboz (Roboz 2005).

It all started with a publication in Lancet in 2002 (Petricoin, Ardekani et al. 2002). The group of Liotta used SELDI to detect proteomic patterns in the serum ovarian cancer patients. The goal was to detect diagnostic biomarkers capable of identifying early-stage ovarian cancer. In the first stage of this project a training set of 50 sera from unaffected women and 50 patients with ovarian cancer were analyzed. Pattern recognition algorithms were used to identify proteomic signatures capable of distinguishing the two groups. In a second round these diagnostic signatures were applied to 116 masked serum samples, which could be classified with a sensitivity of 100%, a specificity of 95% with and a 5% false positive rate. Importantly, all the 50 ovarian cancer serum samples were correctly identified. Although heavily criticized in the scientific community, this paper set the stage for many applications and improvements of the SELDI-TOF-MS methodology in cancer diagnostics. To date there are several hundred papers published applying SELDI-MS to various diagnostic problems.

Laser capture microdissection

Modern proteomics and genomics technology are heavily dependent on technological developments. New microanalytical procedures and instruments are constantly developed and improved to keep up with the increasing demands of the biological sciences. Especially data generated by proteomics depends strongly on sample preparation techniques. The isolation of homogeneous cell population from solid tissues has been problematic. An innovative tool termed "laser capture microdissection" (LCM) was invented at the NIH by Emmert-Buck and colleagues (Emmert-Buck, Bonner et al. 1996). The first commercial product was brought on the market in 1997 through collaboration with Arcturus Engineering Inc.

In principle, LCM instruments are inverted microscopes fitted with a low-power laser. Tissue sections are placed on glass slides and a transparent ethylene-vinyl acetate film will be used to cover the section. The laser will be directed through the transparent film at cells of interest. The ethylene-vinyl acetate film serves several functions:

1. It will adsorb most of the thermal energy and protect the biological macromolecules from damage.
2. The laser will have enough energy to melt the plastic film in precise locations, binding it to cells of interest.

After selection of individual cells on interest, the film is removed together with the adsorbent cells. Cells are now subjected to appropriate extraction and down stream analysis methodologies (e.g. microarray or proteomics). LCM is compatible with several common tissue preparation and staining procedures, although one has to carefully evaluate as these procedures may affect the downstream proteomics techniques. For example, aldehyde-based tissue fixation is known to introduce covalent cross-linking of biomolecules. This could negatively affect protein identification by mass spectrometry. Some of the major disadvantages of LCM are the extremely small amount of isolated sample material, which clearly limits proteomic identification. Additionally, the procedure is very time consuming, especially if larger amount of cells are required.

In conclusion, we believe that LCM will continuously grow to become a major sample preparation strategy for proteomics analysis. Especially improvements in sample preparation and MS technologies will have positive effects on the methodology. We believe that especially in cancer biology and developmental biology LCM-proteomics will produce many novel and exiting results over the next couple of years. We encourage the readers to consult the many excellent papers published on LCM-proteomics in recent years (Ivanov, Govorun et al. 2006).

Molecular imaging of tissue section by mass spectrometry

Direct mass spectrometric analysis of tissue sections is a relatively novel, but very promising technology. To some degree it could almost be considered as a conceptual combination of SELDI and LCM analysis, as molecular mass maps of peptides/proteins are recorded for discrete regions within a tissue section. This technology is powerful enough to systematically detect several hundred polypeptides over a wide mass range (2000-70000 kDa). Statistical and computational analyses of the recorded mass maps have demonstrated the usefulness of this technique for the identification of diagnostic protein patterns (Chaurand, Schwartz et al. 2002; Caldwell and Caprioli 2005; Reyzer and Caprioli 2005).

Briefly, frozen tissue is cut with a cryostat to fine sections (~15µm) and directly applied to a MALDI target plate. Depending on the exact application the target plate is either metal or conductive glass. Glass plates have the advantage that histological staining and visual inspection by trained pathologist could be combined with MS profiling. After drying the tissue section in a desicator, matrix solution is directly applied to the sample. Depending on which type of MS experiment is performed (profiling or imaging) the matrix solution is either applied to defined spots on the tissue or homogeneously distributed over the entire section. MALDI-TOF mass spectra are then directly recorded from this tissue-matrix co-crystal, where each recorded m/z value represents a distinct peptide/protein. Although, it should be noted that unambiguous protein identification cannot be achieved by molecular mass alone. Additional biochemical analysis schemes are generally required. The recent introduction of TOF/TOF mass analyzers capable of generating tandem mass spectra and sequence information directly from some of the recorded peptide peaks will overcome some of these limitations. The continuous improvement of MS hardware will further improve on the basic concept of molecular imaging by MS. A very interesting development in this field is the generation of three dimensional images. The correlation of protein localization obtained from direct tissue MS with anatomical structures of a given tissue could be a very powerful tool. Basically, consecutive sections from a given tissue (in the range of several hundred) are individually analyzed by imaging MS as described above. Computationally extensive algorithms are used to reconstruct a three dimensional image based on these individual sections.

For more detailed information on direct tissue MS, a technique spearheaded by Richard Caprioli from Vanderbilt University, a large number of review and primary literature are available to the interested biologist (Chaurand, Cornett et al. 2006; Meistermann, Norris et al. 2006).

Protein profiling by LC-MS

LC-MS based profiling has developed into the "gold-standard" for the large-scale qualitative and quantitative analysis of proteins. As described above it has several significant advantages over gel-based methodologies (e.g. more comprehensive detection depth, less biased against membrane proteins and proteins with extremes in isoelectric point and molecular mass). A very large number of papers have been published applying LC-MS based proteomics to various questions of cancer biology. Below, we will review some of the major concepts and application.

Proteomics screening of blood samples

One of the major goals of proteome research is the discovery of diagnostic biomarkers. A biomarker is a molecule (e.g. protein, lipid, metabolite etc.) that can indicate a particular disease state. Especially body fluids such as blood, plasma, serum and urine are thought to be excellent sources for the discovery of biomarkers, especially due to the easy, noninvasive availability. As every tissue in the body is perfused by blood, proteins can actively or passively enter the circulatory system. Thus, generating a signature of body fluids containing biomolecules could reflect the ongoing physiological state of a tissue. The body fluid proteome is a very complex mixture, to further complicate the matter a handful of high abundance proteins (e.g. albumin, immunoglobulines) make up for over 80% of the total protein. Although, the low abundance proteins are likely the interesting candidates for biomarker discovery. Finding these biomarkers is like searching for a needle in a haystack. Several analytical strategies have been presented in the scientific literature aiming at reduction of sample complexity with the goal of identifying lower abundance proteins. Many commercial applications are available for the selective removal of high abundance proteins from blood. Although, biologists should keep in mind that many low abundance proteins are transported in the blood by binding to high abundance carrier proteins. By removing the high abundance proteins from the blood many of the putative biomarkers, especially those hatching a ride, might be removed at the same time.

In our opinion the more promising analytical strategy is extensive biochemical fractionation of blood (e.g. ion exchange, size-exclusion chromatography etc.). By reducing the sample complexity of each collected fraction more proteins are identified. One significant disadvantage of this strategy is that the overall analysis time is significantly increased. In other words, detection depth (number of detected proteins) and sample throughput (number of analyzed samples) are two independent parameters of every proteomics analysis.

There have been several large scale human plasma/serum proteomics papers published in the last couple of years. Major datasets come from the group of Leigh Anderson from the Plasma Proteome Institute and from a consortium of several laboratories around the world collectively assembled in the HUPO plasma proteome project. In 2004 Anderson et al. reported a compendium of 1175 nonredundant proteins complied by a combination of literature searches and different separation technologies followed by MS analysis (Anderson, Polanski et al. 2004). The overlap of these four analysis strategies was surprisingly low with only 46 proteins found in each dataset. This clearly highlights the need for multiple independent proteomic platforms for the analysis of plasma samples and the discovery of putative biomarkers.

These known analytical challenges were taken in consideration when the HUPO initiated the Plasma Proteome Project in 2002. A total of 35 collaborating laboratories in multiple countries applied a multitude of analytical technologies to generate a comprehensive, publicly available knowledge base. In the pilot phase of this HUPO project the following points were chosen to be priorities (Omenn, States et al. 2005; States, Omenn et al. 2006):

- The main issues were to test the advantages and limitations of high abundance depletion technologies, fractionation of plasma and the use of different MS technologies.
- Compare reference specimens → serum vs. plasma and the use of different anti-coagulants (EDTA, heparin, citrate).
- Generate a publicly available database.
- Raw data analysis and choice of search algorithm
- Antibody arrays
- SELDI-MS

The analysis of the entire MS/MS datasets against the IPI human protein sequence database revealed 9504 IPI proteins with more than 1 unique peptide and 3020 with more than 2 peptides (Omenn, States et al. 2005). Although, since the presentation of this dataset at the 5th international HUPO conference in Munich, Germany re-analysis of the data with more statistically rigorous parameters suggest the confident identification of only 889 unique proteins. This clearly illustrates the difficulty in validating the large number of putative protein identification obtained in modern profiling experiments, to minimize the false discovery rate (States, Omenn et al. 2006).

Several useful pattern recognition algorithms have been developed in recent years that graphically display LC-MS based profiling results. The groups of Aebersold form the Institute of Systems Biology and Emili from the University of Toronto have presented visualization and alignment tools of LC-MS generated peptide features (Radulovic, Jelveh et al. 2004; Li, Yi et al. 2005). Basically, three dimensional blots, containing the retention time, m/z value and ion intensity of every eluting peptide. Correction and alignment tools are capable of comparing large numbers of these virtual peptide mass maps and identify putative biomarkers. Especially the use of high resolution and mass accuracy mass spectrometry in combination with good and reproducible microcapillary chromatography will further enhance the use of these software tool for biomarker discovery.

Conclusions

In summary, we believe that the constant improvement and technical innovations in proteome research will produce many exiting and unexpected results in cancer biology in the next decade. This will include the discovery of better biomarkers capable of assisting conventional medical diagnostics in an earlier detection of cancer. We also believe that the systematic analysis of mouse model systems of human cancers will lead to a better understanding of the fundamental biological progresses of cancer biology. The better understanding of the molecular and cellular mechanisms could result in the development of better or more specific therapies to finally defeat this horrible disease.

Part III – Computational Platforms

Bill Wong and Igor Jurisica

Addressing important clinical questions in cancer research will benefit from expanding computational biology. There is a great need to support systematic knowledge management and mining of the large amount of information to improve prevention, early diagnosis, cancer classification, prognostics and treatment planning, and to discover useful patterns.

Understanding normal and disease states of any organism requires integrated and systematic approach. We still lack understanding, and we are ramping up technologies to produce vast amounts of genomic and proteomic data. This provides both the opportunity and a challenge. No single database or algorithm will be successful at solving complex analytical problems. Thus, we need to integrate different tools and approaches, multiple single data type repositories, and repositories comprising diverse data types.

Knowledge management is concerned with the representation, organization, acquisition, creation, use and evolution of knowledge in its many forms. Effectively managing biological knowledge requires efficient representation schemas, flexible and scalable retrieval algorithms, robust and accurate analysis approaches and reasoning systems. We will discuss examples of how certain representation schemes support efficient retrieval and analysis, how the annotation and system integration can be supported using shareable and reusable ontologies, and how to manage tacit human knowledge.

Data from high-throughput studies of gene and protein expression profiles, protein-protein interactions, single nucleotide polymorphism, and mutant phenotypes are rapidly accumulating. Diverse statistical, machine learning and data mining approaches have analyzed each of the areas separately. The challenge is to use novel approaches that efficiently and effectively integrate and subsequently mine, visualize and interpret these various levels of information in a systematic and integrated fashion. Such strategies are necessary to model the biological questions posed by complex phenotypes, typically found in human disease such as cancer. Integration of data from multiple high-throughput methods is a critical component of approaches to understanding the molecular basis of normal organism function and disease.

Informatics

Bill Wong

If the 20th century was the age of physics, then the 21st century promises to be the age of biology. New and fundamental advances in genetic manipulation, biochemistry, and bio-engineering are now for the first time allowing us to understand and manipulate, although still in a very primitive way, some of the most intimate biological machinery. In this context, computers are increasingly becoming fundamental mining and discovery tools. We expect that, over the next 10 years, computers running new breed of algorithms – still largely unavailable at the moment – will help automate a significant portion of the drug and diagnostic tool manufacturing process. Moreover, advances in life sciences computational techniques will directly impact a number of other related sectors, from agrochemical research, to bio-engineered products, to polymers and smart materials.

Challenges in the Life Sciences industry

The life sciences arena is experiencing increasing costs, delays, and limitations in the ability to share data as well as the challenges in effectively leveraging the growing volume of data. The following section details the various challenges.

Clinical trials

Current clinical development research is significantly hindered by the lack of available information resulting from the limited interoperability or effectiveness that the current systems have. For instance:

- Test results are often redundantly entered into multiple systems.
- Searching for data is difficult and time consuming because it either can't be accessed or there are wide variations in terminologies and vocabularies used.
- All of this leads to increased costs and inefficiencies.
- Submission of clinical trial data to the Federal Drug Administration (FDA) was paper-based, but the FDA is just now beginning to accept electronic standards-based submissions.
- The clinical trial environment brings a multitude of systems, poor site infrastructure, and a need for industry standards.

These issues result in slow, labor-intense processes that are both expensive and potentially inaccurate.

Without standards, accessing real-time, accurate trial data is difficult. This is due in part to the large volumes of documents that remain in a paper-based environment. Additionally, the lack of an integrated view of clinical, lab, and safety data for clinical research organizations as well as the trial sponsor (a company or organization that provides funding and resources for a clinical trial) leads to delays and potential errors. With limited connectivity between systems, selection of patients for trials must be determined without the aid of computer systems. Furthermore, delays in submission and approval are commonplace.

The increasingly rigorous regulatory requirements cause additional complications. Examples of this added complexity are the stringent FDA requirements for electronic data submissions such as complying with 21 CFR Part 11 and Good Clinical Practice (GCP) requirements. Increased cost due to more complex trial requirements (for example, more patients per trial, the need to cover multiple populations and groups of people with similar characteristics) further reinforce the need for interconnected systems and standards.

Additionally, managing adverse event reporting and communicating the results becomes a slow and cumbersome process that may not provide necessary information in time to protect participants.

Discovery

The limitations in discovery have shifted from data collection to data integration and analysis. Poor data integration is a key factor reducing productivity in research and discovery. The discovery area produces large amounts of data by nature. While storing the large amounts of data is a hardware issue, the real problem is integrating, mining, and analyzing the information.

Another common issue within many pharmaceutical companies is the occurrence of data silos. While this may be a result of internal business processes, data is often unavailable to other researchers in other departments or organizations. This data isolation inhibits advances in the industry.

Communication of standard practices to physicians can take as much as 15 years. Furthermore, data is often stored in different formats and uses special vocabularies to define information. This often leads to manual processing where information must be entered by hand into multiple systems since no connectivity or application integration is present.

Standards

Standards have emerged or are emerging in the healthcare, clinical development, and discovery areas to address the pain points and inefficiencies. Most of these standards are specific to a particular area of the industry, but many address issues that are common across healthcare and life sciences.

Regardless of which branch of the healthcare and life sciences industry they address, there are three main types of standards:

1. Vocabularies and terminologies (often called ontologies) that provide a consistent way to identify and describe things so that data can be readily searched, exchanged, and compared.
2. Data format and messaging standards that specify how to exchange domain-specific data, documents or objects, how to interpret data, and how to report results. By standardizing the format, information can be exchanged in a way that ensures that all parties understand exactly what is meant.
3. Regulations and national initiatives that drive the adoption of existing standards and often the development of new standards required to support the regulation or initiative.

Healthcare standards are not new. They have been evolving for 15 years at organizations such as Health Level Seven (HL7), and therefore are more mature and more pervasive globally within the industry.

Note that governments are increasingly influencing healthcare standards with regulations and initiatives such as proposed electronic health records. In the U.S., FDA regulations regarding the electronic submission of clinical trial data are another example. Fundamental to this government interest is the idea that utilizing standards improves public safety and welfare, reduces opportunities for terrorism, and helps to control customer healthcare costs.

Unlike healthcare, there are few widely accepted standards in the research and discovery areas of life sciences, but there are various emerging standards. For example, genomic content in various public data sources is being standardized. Similarly, there has been major emphasis in life sciences on standards and ontologies for functional genomics. Various groups are attempting to bring together a number of independent efforts in developing controlled vocabularies in the biomedical domain.

Emerging clinical development standards such as those from Clinical Data Interchange Standards Consortium (CDISC) and HL7 will allow electronic submission of clinical trial data. As a result, organizations like the FDA in the U.S. will be able to review the data and approve drugs for use more quickly than with the current paper-based review process.

Several innovations have opened up new opportunities to bring together the worlds of health care and clinical research standards and technology. For example, the integration of XML in healthcare technology solutions is making possible the widespread exchange of information. CDISC, HL7, and the FDA have developed XML-based standards that will eliminate the barrier between data and documents. Both CDISC and HL7's Regulated Clinical Research Management (RCRIM) committee's standards-based information is used to substantiate to the FDA. Other standards development organizations are also working together, realizing the common goal of improving technology within the industry.

Vocabularies and terminologies in healthcare and life sciences

Without terminology standards, health data is non-comparable and cannot readily be searched or accessed. Health systems cannot interchange data, research in a clinical setting is difficult, and linkage to decision support resources is very inefficient.

For interoperability, a system needs common message syntax as well as common vocabularies. These are some examples of vocabularies used in healthcare standards.

- Current Procedural Terminology (CPT) codes—five-digit numbers used to represent medical and psychiatric services given to patients. They are revised each year to reflect advances in medical technology. The 2002 revision contained 8,107 codes and descriptors.
- International Classification of Disease (ICD) codes—a detailed description of known diseases and injuries. ICD-9 and ICD-10 codes are used for inpatient procedures. ICD-9 is used in the USA and ICD-10 codes, which are newer, are used in Europe.

- Logical Observation Identifiers Names and Codes (LOINC)—provide standard codes and nomenclature to identify laboratory and clinical terms and can be used in various other standards.
- Systematized Nomenclature of Medicine (SNOMED)—a dynamic clinical healthcare classification system for the coding of several aspects of a diagnosis.

The National Library of Medicine has the Unified Medical Language System (UMLS) metathesaurus which incorporates medical subject heading (MeSH), parts of ICD, CPT and SNOMED. Additionally, it has more than one million concepts, 5.6 million term names, and greater than 100 source vocabularies.

The U.S. Federal government has come up with recommended terminologies called the CHI (consolidated health informatics), which is a terminology subset. The Federal government also uses these standardized vocabularies internally for example at Veteran's Hospitals. They recommend using:

- LOINC for clinical laboratory results and test orders.
- HL7 vocabulary standards for demographic information, units of measure, and clinical encounters.
- SNOMED CT for laboratory result contents, diagnosis, and problems.
- RxNORM—A set of federal terminologies for describing clinical drugs.
- The National Drug File Reference Terminology (NDF-RT) for specific drug classification.
- EPA substance registry system for non-medicinal chemicals.
- HIPAA transactions and code sets for billing or administrative functions (ICD-9-CM, National Drug Code). The International Statistical Classification of Diseases and Related Health Problems (known as ICD) is a century-old set of heritage and morbidity codes. Since it is mainly for billing, there is uneven granularity. HIPAA mainly references ICD-9-Clinical Modification (ICD-9-CM).

In the genomics and bioinformatics area, naming of the sequences, genes, single nucleotide polymorphisms (SNP), and proteins is very confusing. Currently for the gene symbols and names, the agreed nomenclature is HUGO. There is the Gene Ontology consortium to define various molecular, biological, and cellular functions. Microarray Gene Expression Data (MGED) ontology develops ontologies related to microarray experiments. HUPO Proteomics Standards Initiatives (PSI) is working with MGED to develop emerging ontologies for proteomics.

Examples of common healthcare standards

HL7 Standards

HL7 is both the organization and the collection of standards specifications developed by the organization. HL7's mission is to provide standards for the exchange, management, and integration of data that support clinical patient care, and the management and delivery of healthcare services by defining the protocol for exchanging clinical data between diverse healthcare information systems. The primary HL7 standards are messaging standards and the Clinical Document Architecture (CDA) standard. These enable interoperability across healthcare and clinical development areas (for example, laboratories, pharmacies, patient care, and public health reporting). These standards also address administrative management functions such as accounting and billing, claims and reimbursement, and patient administration. HL7 is also developing the Electronic Health Record System (EHR-S) standard to provide a common language for the healthcare provider community to guide their planning, acquisition, and transition to electronic systems.

HL7 Messaging Specifications

The most widely used of the HL7 standards is a messaging standard (version 2.x, or V2), which allows different health care software applications to communicate with each other. HL7 V2 has both a newer XML version and a non-XML version (most V2 is non-XML). HL7 CDA (clinical document architecture) uses XML.

HL7 Version 2 (V2) is the predominant standard for the exchange of hospital information today because of the early adoption of this standard within the industry. Because of the use of a "structureless" ASCII format; however, it allowed many options and lead to interoperability problems.

HL7 Version 3 (V3) is XML-based and is currently a draft standard for trial use. V3 limits the options and increases interoperability due to a common XML structure. However, making the transition to V3, while beneficial to the industry, will be a major task due to the more complex code structure. Newer healthcare IT projects at HL7 are using V3 already, especially in the area of clinical trials and clinical genomics, since there are few systems using the V2 standard in these areas.

HL7 Clinical Document Architecture (CDA) specification

The HL7 CDA standard provides an exchange model for clinical documents and brings the electronic medical record within reach for the healthcare industry. Published by HL7 in October, 2000 and approved by the American National Standards Institute (ANSI) in November of the same year, CDA was the first XML-based standard for healthcare. CDA (previously known as Patient Record Architecture) provides an XML-based exchange model for clinical documents such as discharge summaries and progress notes. Because of the use of XML, both humans and machines can read and process CDA documents. XML-enabled Web browsers or wireless devices such as cell phones and PDAs can also display CDA documents, making information available to a physician or others who may be miles away from a patient.

Electronic Health Record (EHR) at HL7

Electronic health records are an essential part of the solution for the healthcare industry. HL7 is leading the development of the Electronic Health Record System (EHR-S) model to provide a common description of the functions in an EHR system. It enables all stakeholders involved in describing EHR-System behavior to have a common understanding of those functions. The standard is designed to accommodate not only inpatient and outpatient care, but also long term care (in a nursing home, for example) and care in the community. Additionally, it is expected to provide a comprehensive set of functional terminology, which will be referenced in the specification of EHR Systems by health care providers, software system suppliers, and system certification authorities.

In 2004, HL7 successfully developed and balloted an Electronic Health Record System Functional Model and Standard as a "Draft Standard for Trial Use" (EHR-S DSTU).

In 2005, HL7 and the Object Management Group (OMG) formed a joint project to create healthcare-related Web services focused on EHR. One example of a Web service would be Master Patient Index service, which is useful to match up patient records from different hospitals and clinics, since each medical institute assigns their own unique patient identifier. Some countries, such as the UK, have a countrywide patient identifier.

Digital Imaging and Communications in Medicine (DICOM)

DICOM standards are used or will soon be used by virtually every medical profession that utilizes images. These include cardiology, dentistry, endoscopy, mammography, ophthalmology, orthopedics, pathology, pediatrics, radiation therapy, radiology, and surgery. DICOM creates and maintains international standards for communication of biomedical diagnostic and therapeutic information in disciplines that use digital images and associated data. The goals of DICOM are to achieve compatibility and to improve workflow efficiency between imaging systems and other information systems in healthcare environments worldwide. These images can be part of an electronic medical record.

Examples of common clinical trial standards

Clinical trial standards are used for sending clinical trial data results electronically instead of by paper and for exchanging information about trial data such as lab test data. Clinical trial standards are used by pharmaceutical companies and labs testing the data. For example, it can take six months and thousands of dollars to set up a proprietary exchange format between a pharmaceutical company and a central lab related to clinical trial lab test data. Using the CDISC lab standard reduces the time involved in setting up a unique data exchange format. Additionally, the central labs that perform tests on the data typically offer pharmaceutical companies a lower cost to use the CDISC standards than to set up yet another unique clinical trial study.

Clinical trial standards are created by two standards organizations that collaborate: HL7's RCRIM committee and CDISC.

Clinical Data Interchange Standards Consortium (CDISC)

The CDISC develops the clinical trial industry standards that support the electronic acquisition, exchange, submission, and archiving of clinical trial data. CDISC standards enable information system interoperability to improve medical research and related areas of healthcare.

- **Submissions Data Standards Team** (SDS) guides the organization, content, and form of submission data for clinical trials.
- **Operational Data Model** (ODM) describes the format for data collected in a clinical trial to facilitate data exchange and archiving. It is the submission standard used in clinical genomic solutions.

- **LAB Model** also known as the **CDISC Laboratory Data Standards Model** describes requirements to improve laboratory data interexchange between pharmaceutical companies and central labs running lab test of the clinical samples. The CDISC Laboratory Data Standards Model is the first step in proposing independent standards for the interchange of clinical trial laboratory data.

HL7 Regulatory Clinical Research Information Management (RCRIM) Committee

CDISC and the HL7 RCRIM committee work together to create clinical trial standards. CDISC models are created and brought to HL7 RCRIM to create HL7 V3 messages. Since HL7 is an ANSI-accredited standards organization, government entities, such as the FDA, are actively participating in HL7 creating clinical trial standards. The long-term goal is to have an XML-based HL7 messaging or CDA format.

Examples of common discovery standards

Discovery standards are emerging to address the sharing of data across the proteomic and genomic research populations in life sciences. Standards are emerging for proteomics research, which is the study of the proteins within a cell, a research field complementary to genomics. Proteomics is of interest to the biotech industry in the research and development of new drugs. The microarray and proteomics standards are XML-based.

Some of the standards in this arena include:

- **MicroArray and Gene Expression** (MAGE) deals with a common data structure for microarray-based gene expression data. It is used to exchange microarray data. This is commonly used when testing and reporting on DNA fragments, antibodies, or proteins.
- **Minimum Information About a Microarray Experiment** (MIAME) is the standard that describes what is needed to enable

the interpretation of the results of the experiment unambiguously, and potentially to reproduce the experiment. MIAME deals with research processes that use MAGE data.
- **Minimum Information about a Proteomics Experiment** (MIAPE) contains guidelines on how to fully report a proteomics experiment.
- **Proteomics Standards Initiative** – Markup Language (PSI-ML) is an XML format for data exchange. Derived from the Global Proteomics Standards (GPS) and the proteomics workflow/data object model, PSI-ML is designed to become the standard format for exchanging data between researchers, and submission to repositories or journals. The model and the exchange format will eventually form part of the emerging functional genomics model, which will be developed by GPS in collaboration with the developers of the MAGE model for transcriptomics.
- **Life Science Identifiers** (LSID) provide a simple and standardized way to identify and access distributed biological data. LSIDs refer to persistent, location-independent resource names. The LSID protocol will enable scientists and researchers across multiple organizations to share data and collaborate. LSIDs are utilized in multiple standards organizations such as MGED, HL7, and others.
- **Secure Access for Everyone** (SAFE) specifies a public-key infrastructure (PKI)-based framework for legally binding and regulatory compliant digital signatures for business-to-business (B2B) and business-to-government (B2G) transactions across the biopharmaceutical community. Global biopharmaceutical companies drive SAFE. These companies see great value in paperwork reduction and improved compliance-readiness when using SAFE-enabled applications. A subset of these companies plan to use the SAFE PKI for additional applications such as physical and logical access control.

Figure 14 presents standards related to research, clinical trials and healthcare in general.

Informatics

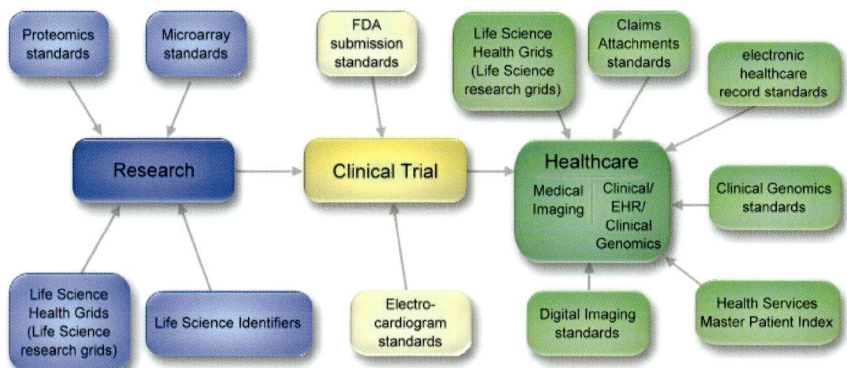

Figure 14. Multiple standards related to research, clinical trials and healthcare in general. Source: IBM.

Regulations and national initiatives

Regulations and national initiatives both drive the development of new standards and the adoption of established standards aimed at regulatory compliance. Communities around the world are recognizing the benefits of interconnected systems and are contributing to the development and adoption of standards. The following are examples of healthcare activities from regions around the world.

European Union

The European Union has established several initiatives for healthcare that reinforce the need for standards. The European Committee for Standardization (Comité Européen de Normalisation, CEN) has already had extensive eHealth-related standards activities. HL7 is also involved in European standards development activities.

European healthcare standards development considerations include:

- Deploying e-Health systems into member European Union (EU) countries.

- Interoperability and the use of electronic health records from country to country with language considerations.
- Reimbursement of e-Health services across the various EU countries to resolve the issue of various payers that include insurance, government, and employers.
- Interoperability standards for health data.
- Deployment of health information networks for e-Health based on fixed and wireless broadband and mobile infrastructures, and Grid technologies.

Australia

In November 1999, the Australian National Health Information Management Advisory Council (NHIMAC) released '*Health Online: A Health Information Action Plan for Australia.*' Health Online was the Australian national strategy for information management and the use of online technologies within the health sector, and also detailed a series of action plans for nationally significant projects. One of the key recommendations in Health Online was the development of a national framework for the use of electronic health records to improve the efficiency, safety, and quality of care compared with paper-based systems.

The Australian Government Department of Health, Ageing and Standards is also participating in CEN working groups to ensure that Australian health care interests are incorporated in any CEN standards development.

China

China's Center for Disease Control now has a system that allows daily updates from 16,000 hospitals nationwide, providing information on 32 different diseases. Another example in China, the city of Foshan is participating in a government experiment with a goal of linking 20 hospitals (with 10,000 beds) in the city, allowing them to exchange medical records with one another electronically. The city has a 10GB network connecting the 20 hospitals and two clinics taking part in the program.

Japan

At the end of 2001, Japan's Ministry of Health, Labor, and Welfare formulated a "grand design for computerization of the medical care field." Targets included the spread of electronic medical record filing systems to at least 60 percent of all hospitals with 400 beds or more, and the spread of computerized medical treatment statements to 70 percent or more of all hospitals in Japan. According to the "e-JapanII" strategy, by 2005 the authentication infrastructure will be put into order, including approval for storing electronic medical records outside medical institutions. Another target, to be achieved by 2010, is to switch to an online system of electronic medical statements for all medical institutions that apply for the change.

Canada

The Canadian eHealth Initiatives Database is a collaboration between the Health and the Information Highway Division, Health Canada, and the Canadian Society of Telehealth. This collaborative effort is a searchable database that profiles Canadian telehealth, electronic health records, education and training, and health information infrastructure initiatives and programs.

The Canadian Health Network (CHN) is a national, bilingual, Internet-based network of health information providers. It provides Canadians with an accessible Internet gateway to information on healthier lifestyles, disease prevention, and self-care from respected Canadian government and non-governmental organizations in a non-commercial format.

The Centre for Surveillance Coordination (CSC) collaborates with public health stakeholders on the development, maintenance, and use of data about cases of nationally notifiable diseases (health surveillance information), tools, and skills that strengthen Canada's capacity for timely and informed decision-making. The Centre aims to increase the capacity of public health professionals and decision-makers across Canada to better protect the health of Canadians.

United States

The Health Insurance Portability and Accountability Act (HIPAA) will ensure privacy and security as health insurance is linked electronically with healthcare systems. HIPAA requires that standards be developed to ensure the security of individually identifiable health care information.

National initiatives like the National Health Information Infrastructure (NHII) in the US are also forming to guide the move to electronic health records. The NHII will concentrate on creating institutions that can set standards for health information technology and helping firms acquire financing for the systems they need. This will set the stage for widespread implementation of EHRs in the U.S. While this is a U.S. initiative, it is addressing a need that is common around the world.

EHR standards are gaining national interest in the U.S. The Health and Human Services Secretary requested HL7 to accelerate the development of the Electronic Health Record-System (EHR-S) standard in 2003. Additionally, in 2004, the National Health Information Technology Coordinator position was created with the goal of making a nationwide EHR system a reality within 10 years. In the U.S., the financial benefit could be as high as $140 billion per year through improved care, reduced duplication of medical tests, and reductions in morbidity and mortality rates.

Regional Health Information Organizations (RHIO) enable the efficient exchange and use of clinical health care information to improve health care quality, safety, and productivity across wide-ranging communities, both geographic and non-geographic. RHIOs exchange patient information within a region or group of hospitals and clinics or HMO hospitals that belong to the same participating systems. Furthermore, RHIOs may become vehicles for administering financial incentives to support IT investment and use. Fueled by federal and private investment, RHIOs are in the early stages of development in communities throughout the U.S.

Certification Commission for Healthcare Information Technology (CCHIT) is a voluntary, private-sector initiative whose purpose is to create an efficient, credible, sustainable mechanism for the certification of healthcare information technology products.

Where is all this technology taking us?

Ultimately, the goal for medicine is to anticipate the need for medical treatment and define treatments that are specific for each person. Open standards, in conjunction with the following three stages, are necessary in order to reach the goal:

- Define and deploy a fully paperless medical record system.
- Build electronic links between and among institutions.
- Link clinical and research databases.

These steps will allow healthcare and life sciences to rapidly evolve. Open standards allow each step to become a reality. The climb to personalized information based care by using information technology is highlighted in Figure 15.

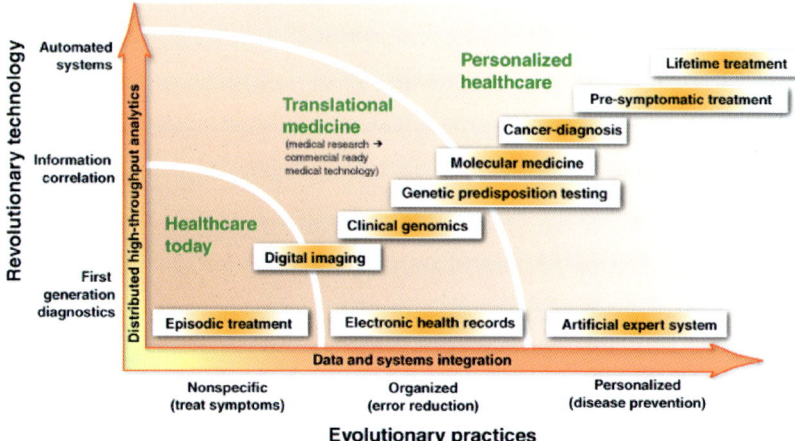

Figure 15. Trend of moving from current healthcare standards toward translational and personalized medicine by integrating information, and automating the diagnostic process. Source: IBM.

Standards Organizations

The following is a brief summary of some the standards organizations that are addressing various problem areas in healthcare IT.

Accredited Standards Committee (ASC X12)

ANSI Accredited Standards Committee (ASC) X12N develops and maintains X12 EDI and XML standards and guidelines. Payors use X12 standards. The insurance subcommittee develops and maintains standards in insurance related business processes such as healthcare. Example X12 healthcare related work groups include: healthcare eligibility, claims related information, interactive claims, transaction coordination and modeling, services review, patient information, provider information, and HIPAA related coordination. X12N and HL7 have a joint project on claims attachments.

The healthcare industry utilizes X12N's standards in transactions with trading partners.

Clinical Data Interchange Standards Consortium (CDISC)

The CDISC develops the industry standards that support the electronic acquisition, exchange, submission, and archiving of clinical trial data.

Digital Imaging and Communications in Medicine (DICOM)

The organization produces standards that aid in the viewing of images plus image-related medical information. Additionally, DICOM creates standards that assist with the associated interoperability between systems that generate and handle patient image and image related information. DICOM standards include digital formats for non-radiology images that are components of a digital patient medical record.

Global Grid Forum (GGF)

GGF is a worldwide, community-driven forum of individual, academic, government, and corporate researchers and practitioners working on distributed computing or grid technologies. In October, 2002, GGF established its first industry- and academic-focused research group, Life Sciences Grid.

The Life Sciences Grid Research Group (LSG-RG) explores issues related to the integration of information technology with the Life Sciences on a grid infrastructure. Some of their goals outlined on their website include: provide clear examples of the diverse use of grid in life sciences, discuss issues of access to data in life sciences, identify how the grid is being challenged by the life sciences and where there is need for activity, and identify different solution areas and possible reference architectures. Projects include researching best practices for health grids and life sciences grids. Newer projects are focused on understanding common practices and security issues in health grids.

Health Level Seven (HL7)

HL7 is recognized as the key information technology standards organization within the international hospital and clinical community. The focus is on producing standards that facilitate the exchange of clinical records, medical procedures and other related information.

Integrating the Healthcare Enterprise (IHE)

IHE is an initiative by healthcare professionals and industry to improve the way computer systems in healthcare share information. Rather than develop individual standards, IHE promotes the coordinated use of established standards such as DICOM and HL7 to address specific clinical needs in support of optimal patient care. IHE publishes profiles that define a blueprint for how existing standards can be applied to address specific scenarios within a number of healthcare domains including radiology, cardiology, laboratory and IT infrastructure. One of the goals of IHE is to foster smoother and less costly deployment of healthcare IT systems by ensuring consistent, cross-vendor support for a core set of standards.

Microarray Gene Expression Data (MGED)

The MGED Society and the Object Management Group (OMG) worked together to create MAGE, a standard for exchanging microarray data generated by functional genomics and proteomics experiments. This is an important standard within the life sciences industry. Enhancements to MAGE are being worked on in both OMG-LSR and MGED.

Proteomics Standards Initiative (PSI)

HUPO PSI is creating standards for proteomics data representation to facilitate data exchange, storage, data comparison and verification.

PSI is developing standards for mass spectrometry, protein-protein interaction data and General Proteomics.

How are these standards helping to make a difference?

Healthcare Collaborative Network

The Healthcare Collaborative Network (HCN) began as an effort to demonstrate how an interconnected, electronic information infrastructure could be used for the secure exchange of healthcare data to enable the detection of and response to adverse healthcare events, including bioterrorism. The project is supported by IBM, the eHealth Initiative, Connecting for Health, and others. Originally, the idea was mainly to provide electronic reporting to the CDC, however it became clear that the same information could be of use across the federal government.

Federal agencies already required detailed reporting from healthcare entities. The demonstration focused on using existing open standards and technologies to enable the electronic reporting of that healthcare data. The HCN architecture utilized a publisher-subscriber basis for the exchange of information. Participants used an Internet portal to indicate the types of information they wanted to receive or make available. Data was transferred using existing open standards (in this case, encrypted HL7 messages wrapped in XML).

Key design elements include:

- HCN uses existing data available in most provider settings (ICD, CPT, LOINC, NDC via HL7) Data review organizations (or subscribers in the publisher-subscriber system) request data Data source organizations (publishers) approve the reviewers' requests for data.
- HCN is compliant with HIPAA regulations and includes strong security measures for authentication and encryption.

- HCN is based on open standards and the approach is nonproprietary.

In addition to the demonstration project, HCN is a long-term strategy to improve healthcare delivery and aid in the rapid detection of and response to adverse healthcare events.

The long-term goals are:

- Create an information network that enables the secure transmission of healthcare data.
- Implement HCN nationally through the healthcare ecosystem.
- Improve the collection, dissemination and analysis of healthcare data.
- Create an infrastructure for the detection of and rapid response to bio-surveillance, adverse healthcare events, and inappropriate care.
- Improve the reporting and analysis of healthcare data.

Potential benefits of the HCN include:

- For patients, the benefits of the national implementation of HCN include a reduction in errors, a higher quality of care, and improved outcomes. HCN also offers the possibility of increased patient participation in meeting their healthcare needs, with the assurance that privacy and security rules are followed.
- For clinicians, the access to patient and healthcare data collected from multiple points throughout the healthcare system can improve their decision-making.
- For healthcare systems, the open standards-based infrastructure can help them improve patient care, reduce the burdens of reporting requirements and lower the costs of integrating systems.
- For healthcare payers, such as insurance companies, the benefits include an improved ability to evaluate and manage the effectiveness and quality of care and lower costs.
- For public health, the benefits include rapid access to critical data that can aid their decision-making and response.
- For pharmaceutical development and clinical researchers, the quicker access to up-to-date data improves the efficiency of accessing critical information.

- For quality improvement organizations, the electronic access to data should reduce the costs related to the accreditation process for all parties. They should also be able to more effectively measure health outcomes.

Mayo Clinic and the clinical genomics solution

Mayo Clinic is one of the world's leading clinical research organizations and no stranger to cutting-edge technology. An early adopter of electronic medical records, one of the goals of Mayo Clinic is information-based medicine. Information-based medicine refers to the practice of taking the results large-scale clinical analysis and leveraging that information to create customized patient treatments based on the specifics of the patient, their conditions and background.

A key factor in medical advances is the circular flow of knowledge between clinical research and clinical practice. The knowledge gained from research and patient care is put to use in clinical practice. In turn, the data generated by patient care spurs clinical research. Current trends are accelerating that flow and promising exciting medical advances.

Always seeking to improve the way it diagnosis and treats illnesses, Mayo Clinic looked at an intriguing combination of trends.

- There has been a huge increase in the volume of clinical data, generated in part by an increase in the number of tests available to doctors and the types of tests.
- Even more data is made available by the steady adoption of electronic medical records. Recent medical breakthroughs in genomics and proteomics hold the potential for advancements in understanding diseases at a molecular level.
- Advances in medical information technology, including the development of powerful tools for integrating systems and the rise of open and industry-specific standards, offer the ability to federate different types of data generated from multiple different sources.

With these developments in mind, Mayo Clinic wanted to build a new infrastructure to tap into the abundance of data held in the 4.4 million electronic patient records it had on hand and take a large step closer to the goal of information-based medicine.

The challenges were daunting. Even with standards and tools available, integrating such a vast amount of data stored in different formats and from different sources is a highly complex task. In addition, with the large number of users for an integrated system and the HIPAA-related patient confidentiality issues, access and use of the system requires powerful security measures and logging abilities. Finally, the Mayo Clinic had to ensure that the system could accommodate new data sources as they arose and be able to export usable information to a range of potential analysis and clinical decision-support tools.

In collaboration with IBM, Mayo Clinic identified the key sources of the data they wanted to work with. This included electronic medical records, lab test results, and billing data, which provided patient demographic information and standardized diagnostic codes.

From there, Mayo Clinic and IBM designed a system using state-of-the-art technologies and open standards that would:

- Handle the storage of a large quantity of data.
- Provide a front-end tool for building queries and returning results.
- Address the necessities of security compliance.
- Allow for additional development and new applications in the future.

The system leverages existing open standards already in use at Mayo Clinic such as MAGE and standards from HL7. IT-specific standards such as Web services, SOAP, and XML provide necessary communications capabilities, and make it possible for Mayo Clinic to link legacy applications to the data warehouse. Making further use of Web services standards, new applications can be deployed as plug-ins.

To enable compliance with HIPAA privacy regulations and meet the security concerns, the solution includes a strong authentication system with the capability to set access rights on a per-person basis. It also includes auditing capabilities that log every action taken by every user on the system.

Potential benefits include:

- The real-time access to clinical, genomic and proteomic data will allow for more targeted and effective treatments, leading to better outcomes and lower costs.
- The open architecture allows Mayo Clinic to easily create and implement new tools for analysis and clinical decision-support.
- The new system drastically reduces the time involved in the task of finding participants for new studies, which in turn accelerates the speed of new research.
- Once Mayo Clinic creates XML-based links to major, external sources of genomic and proteomic data, such as the National Cancer Institute, it will be able to reduce the time involved in Clinical Genomics (the correlation of genetic data with data on treatment effectiveness) from months or years to a matter of minutes. Combined with the ability to create patient profiles at the genetic level, Mayo Clinic will be able to create highly targeted treatments.

Drug Discovery is an Information Technology Problem

Today, increasingly complex models and growing amounts of data are intensifying the need for increased memory and compute power, now more than ever. Data complexity is growing faster than it can be absorbed with traditional methods. And, it is becoming more common for important jobs to run for ever longer periods of time, putting additional demand on computing resources. A new approach is needed that can offer high performance and extreme scalability in an efficient, affordable package that provides a familiar environment to the user community.

BlueGene/Light (BG/L) is a highly parallel supercomputer, consisting of up to 64K nodes, that IBM is building with partial funding from the U.S. Department of Energy. It uses system-on-a-chip technology to integrate both computing and networking functions onto a single chip (see Figure 16). This high level of integration and low power consumption permits very dense packaging: 1,024 nodes (representing up to 5.6 Teraflops) can be housed in a single rack. We describe the two primary BG/L interconnection networks: a torus network for point-point messaging, and a tree network for I/O, global reductions and broadcast.

Blue Gene/L Scalability

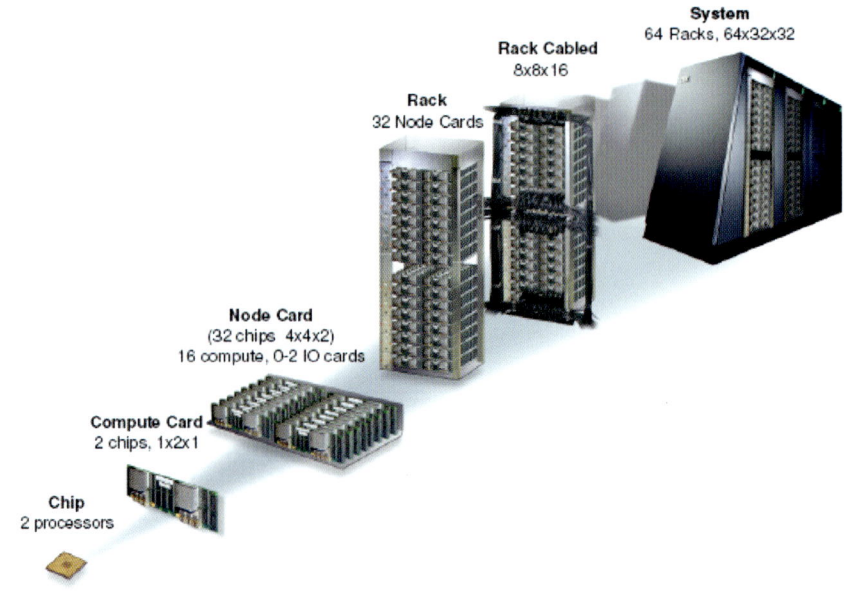

Figure 16. BlueGene scalability. BlueGene/Light integrates both computing and networking functions onto a single chip. The high level of integration and low power consumption permits dense packaging – 1,024 nodes (representing up to 5.6 Teraflops) can be housed in a single rack, and 64 racks can be integrated into one system. Source: IBM.

Blue Matter is the parallel software framework for performing molecular dynamics simulations on BG/L. It has many novel design features tailored to investigations of scalability on the thousands of nodes that BG/L provides. A key goal of the Blue Gene project has been to increase the understanding of protein science using large-scale simulation, and results of recent simulations on systems ranging from small peptides in water to large proteins in a lipid membrane will be shown. The current system running in 'production' mode is the light receptor, rhodopsin, in a lipid/cholesterol bilayer similar to a cell wall. The simulation contains 43K atoms, and is running on 512 nodes of BG/L.

Information-Based Research

Many researchers and academic institutions have adopted the use of open source databases, such as MySQL or PostgreSQL. Research departments from the large pharmaceuticals will often use the commercial information management solutions from IBM and Oracle. However, the technology alone will not suffice, it is essential to follow good principles:

- Capture and store actual content – rather than just images – of printed reports generated by instruments for scientist review, using a database.
- Create report summaries, presentations, electronic submissions and publications – for example, database technologies can address the need to store and integrate data from molecular profiles with clinical and path information and then submit this to an online repository such as GEO or ArrayExpress while conforming to the MIAME standard.
- Dynamically link instrument files and interpreted data to collaborate electronically with colleagues.
- Catalog data according to appropriate protocols and projects with no manual intervention by analysts.
- Secure one enterprise-wide catalog and archive strategy for your lab instruments and business applications.
- Search across locations, databases and projects around the world, and integrate heterogeneous data in multiple formats.
- View multiple reports from disparate sources simultaneously.

- Access data for the long-term, without the instrument software that created it.
- Select and send data to various applications and programs for summary reports and scientific collaboration.

Analysts and lab managers can work with confidence knowing that data is safely and securely archived, and can be easily accessed when required.

The volume of data for researchers involved in life sciences can double every six months. This rate of growth exceed Moore's law, which predicts that capacity doubles every eighteen months (see Figure 17).

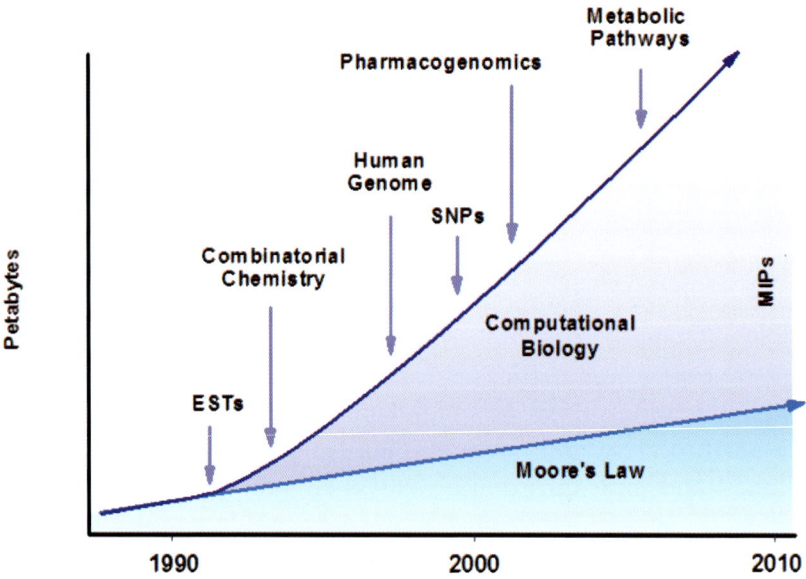

Figure 17. Growth of computational needs in biomedical field, as compared to the Moore's Law.

A data warehouse is an application used to collect and manage data from various data sources. The data is imported from the source applications, stored centrally, and further processed to fit the needs of the end users. The main characteristics of a data warehouse include:

- Large amounts of data.
- Data enters data warehouse via ETL programs.
- Not a transactional database.
- Typically used to support BI applications.

A data warehouse is a central repository of summarized data from disparate internal operational systems and external sources. Operational and external source data is extracted, integrated, summarized, and stored into a data warehouse which can then be accessed by users in a consistent and subject oriented format. Being organized around a business entity such as customer, product, or geographical region, is more useful for analysis, as opposed to applications, which tend to be designed to support a vertical function of the business such as order-entry, accounts receivable or general ledger.

A data warehouse has a very different structure compared to an operational transaction-based system. Data may be:

- Archived and summarized as opposed to current.
- Organized by subject as opposed to application.
- Static until refreshed as opposed to dynamic.
- Simplified for analysis as opposed to complex for computation.
- Accessed and manipulated as opposed to updated.
- Unstructured for analysis as opposed to structured for repetitive processing.

The data warehouse overcomes limitations of decision-support systems:

- Complex *ad-hoc* queries are submitted and executed rapidly because the data is stored in a consistent format.
- Queries don't interfere with ongoing operations because the system is dedicated to serving as a data warehouse.
- Data can be organized by useful categories such as customer or product because the data is consolidated from multiple sources.

In short, the data warehouse is a single source of consolidated data, which provides an enterprise-wide view of the business.

Federated Data Access

BLAST (Basic Local Alignment Search Tool) is probably the single most-used algorithm in bioinformatics research. It requires nucleotide or protein sequences to initiate the search, as well as various search parameter values used to fine-tune the specifics of each search, and returns the sequence "hits" that are most similar to the input sequence based upon specialized search and comparison algorithms intrinsic to BLAST. The most well-known and frequently used BLAST search tool is available from NCBI (blastall); however, there are other variants of BLAST, such as the TurboBLAST® blast accelerator from TurboWorx®, Inc.

Researchers frequently wish to integrate the BLAST algorithm with other data sources, either to supply BLAST query sequences or to provide additional annotations on sequences that are found to match.

The Online Predicted Human Interaction Database (OPHID; http://ophid.utoronto.ca) is an online database of human-protein interactions (Brown and Jurisica 2005). OPHID has been built by combining known interactions, with interactions from high-throughput experiments, and interactions mapped from high-throughput model organism data to human proteins. Thus, until experimentally verified, these "interologs" (i.e. interactions predicted by using model organism interactions between interologous proteins) are considered predictions. Since OPHID supports batch processing in multiple formats (Figure 18), it can facilitate interpretation of microarray experiments and other integrative data analysis (Barrios-Rodiles, Brown et al. 2005; Brierley, Marchington et al. 2006; Kislinger and Jurisica 2006; Seiden-Long, Brown et al. 2006; Motamed-Khorasani, Jurisica et al. 2007). For model organism studies (e.g. rat, mouse, fly, worm, yeast), there is an analogous repository of known and predicted interactions, I2D (Interologous Interaction Database; http://ophid.utoronto.ca/i2d), which comprises 337,712 interactions, including 182,105 source and 158,620 predicted interactions.

To help reduce false positive interaction predictions it is useful to integrate multiple sources of supporting evidence, such as colocalization, interaction domains, sequence identify for orthologues, co-expression, etc., In addition, an automated text mining system may help to find relevant literature by analyzing PubMed (Hoffmann and Valencia 2005; Otasek, Brown et al. 2006). The goal is to automatically identify abstracts that provide positive (or negative) evidence for an interacting protein pair. The main challenge is to unambiguously identify protein names, and evidence for interaction. The process requires several steps, as outlined in Figure 19.

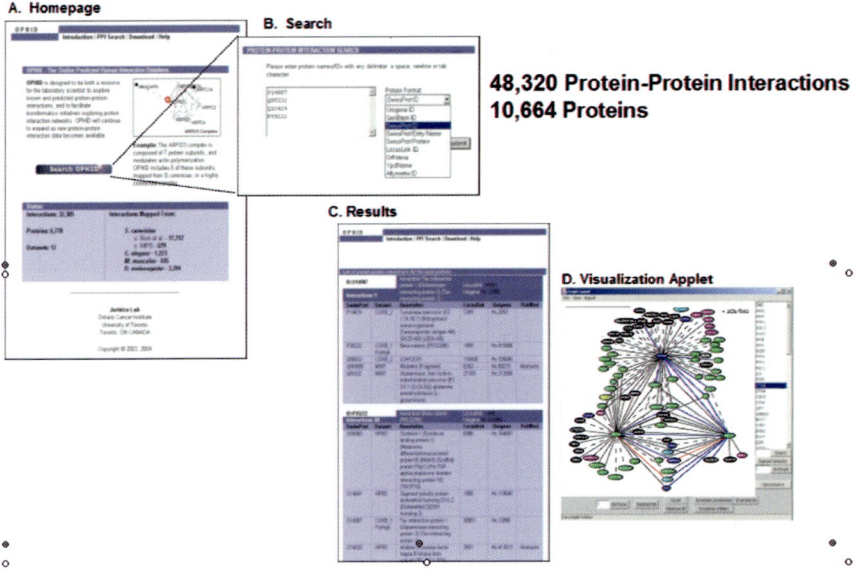

Figure 18. OPHID protein-protein interaction web resource. http://ophid.utoronto.ca. Interactions can be searched in a batch mode using multiple identifiers (SwissProt, Unigene, Locuslink, etc.). The results are displayed in html, ASCII-delimited or PSI (Hermjakob, Montecchi-Palazzi et al. 2004) formats, or graphically, using NAViGaTor (http://ophid.utoronto.ca/navigator).

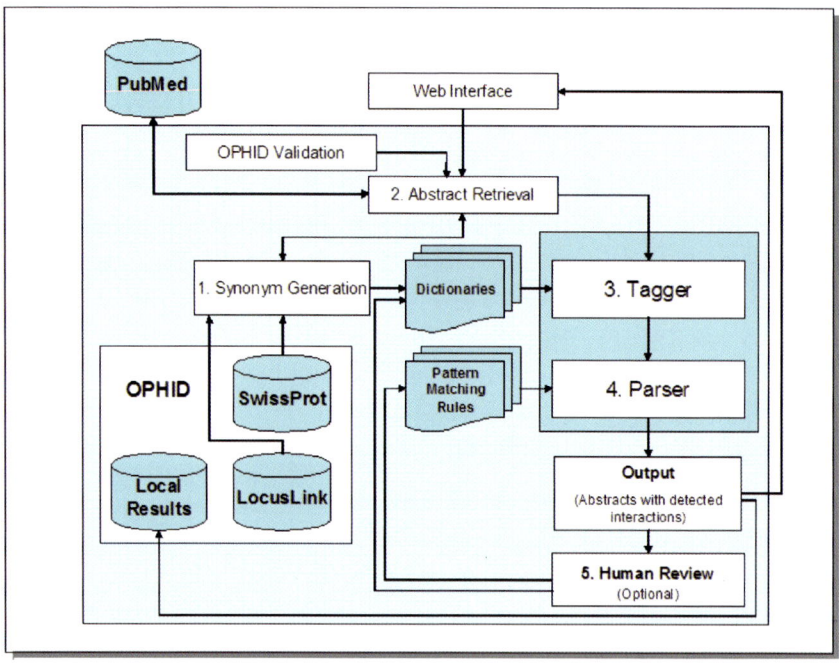

Figure 19. Automated validation of predicted interaction using text mining of PubMed abstracts (Otasek, Brown et al. 2006).

Additional challenge of online resources that support integrative computational analysis is scalability and flexibility. Addressing these issues requires a database integration middleware.

Using, database integration middleware such as IBM WebSphere Information Integrator™ — a robust, user-friendly middleware technology that provides integrated, real-time access to diverse data as if it were a single database, regardless of where it resides (see Figure 20). Wrappers can be used to expand the data types that can be accessed through WebSphere Information Integrator, some examples include:

- Entrez—direct and fast access to key Pubmed, Nucleotide and Genbank data sources.
- Blast—more power for gene and protein similarity searches.

- HMMER—an SQL-based front end to the HMMER application.
- XML—SQL-based access to XML-based data sources.
- BioRS—access to a broad array of public bioinformatics data sources.
- ODBC—access to additional relational data sources.
- Extended search—integrates information from unstructured data sources.

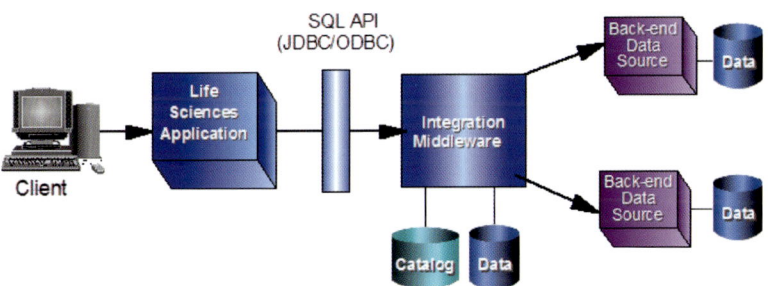

Figure 20. Middleware for life sciences: WebSphere Information Integrator.

Once a virtual database is set up with WebSphere Information Integrator, labor-intensive, repetitive and error-prone probes heterogeneous data sources can be eliminated.

Scenario 1: Given a search sequence, search nucleotide (NT), and return the hits for only those sequences not associated with a Cloning Vector. For each hit, display the Cluster ID and Title from Unigene, in addition to the Accession Number and E-Value. Only show the top 5 hits, based on the ones with the lowest E-values. 2

    Select   nt.GB_ACC_NUM,   nt.DESCRIPTION,   nt.E_VALUE, useq.CLUSTER_ID, ugen.TITLE
    From ncbi.BLASTN_NT nt, unigene.SEQUENCE useq, unigene.GENERAL ugen
    Where   BLASTSEQ   = 'GGCCGGGCGCGGTGGCTCACGCCTGTAATCCCAGCACTTTGGGAGGC
    CGAGGCGGGCGGATCACGAGGTCAGGAGATCGAGACCATCCTGGCTAACACGGTGAAACCCCGTC'
    And nt.DESCRIPTION not like '%cloning vector%'
    And nt.GB_ACC_NUM = useq.ACC
    And useq.CLUSTER_ID = ugen.CLUSTER_ID

Order by E_VALUE FETCH FIRST 5 ROWS ONLY

<u>Scenario 2:</u> Return only BLAST alignments in which the subject sequence contains the P-loop ATPase domain [GA]xxx GK[ST]. This query uses the LSPatternMatch( ) function to filter BLAST results using Perl regular expression syntax, which allows more powerful pattern matching than the traditional SQL LIKE statement. This specialized function is one of several that IBM includes within DB2 II as part of its set of Life Sciences User Defined Functions (UDFs).

```
Select a.gene_id, b.accession_number
From myseqs a, myblastp b
Where b.Blastseq=a.sequence
And LSPatternMatch(b.HSP_H_Seq, '[GA].{3}GK[ST]') > 0
```

Another advantage of a federated data model such as WebSphere Information Integrator is the ability to have a "metadata" view of the enterprise. For example, WebSphere Information Integrator can be used to register all the diverse data and formats (relational non-relational/unstructured) in the organization – across all geographies, departments and networks. While robust security is in place under such a federated data model (i.e. users can only see data consistent with their privileges), scientist or administrator needs to go to one place to see where all of the data across the enterprise is located, and format it is in. Furthermore, changes at the data sources may be incorporated into the federated data model by updating the nickname configurations, helping to eliminate the need to modify applications due to data source changes.

Online Analytical Processing (OLAP)

Online Analytical Processing (OLAP) enables multidimensional analysis of data. OLAP servers are multidimensional analysis tools that enables these star schema relationships to be stored in relational format, or within a multidimensional format, for greater performance. In the example above, the dimensions (Time, Patient data, Species) can have a hierarchical structure. This is very good from a multidimensional organization point of view (see Figure 21).

Biological data is very complex and inter-linked. There are many hierarchies of information of functions at the structural, cellular and molecular levels built on top of information coded in the genes. On a very general level, the goals of molecular biology are to identify the genes within an organism and match the proteins that they code for, understand the function of the individual proteins and then understand how proteins function together. This hierarchical nature or characteristic biological data makes it a natural fit for OLAP technology. OLAP technology allows researchers to analyze the data at all scales of biological relevance and navigate through multidimensional hierarchies and understand relationships faster.

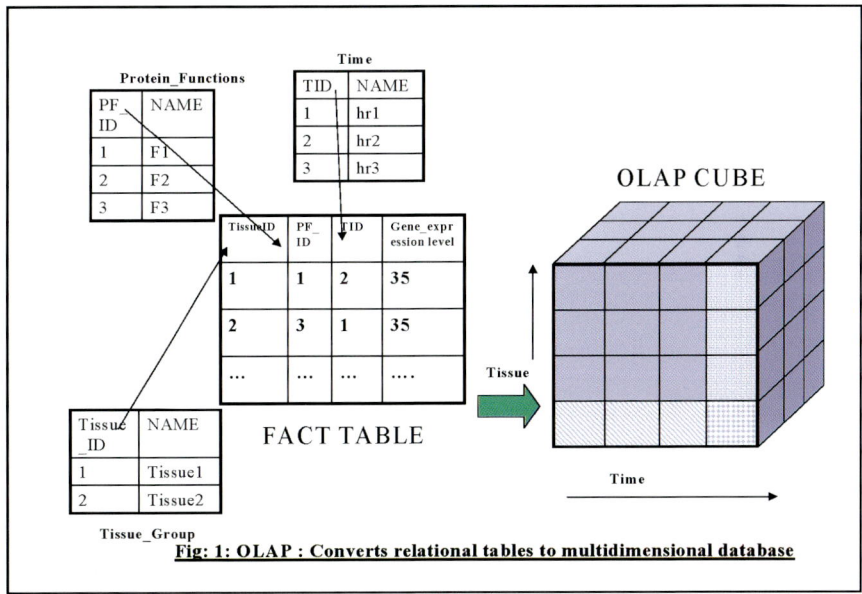

Fig: 1: OLAP : Converts relational tables to multidimensional database

Figure 21. OLAP – converts relational tables to multi-dimensional database.

The following are scenarios of how OLAP technology can be used:

- **Clinical Trial Analysis and Tracking** – To demonstrate drug efficacy and side effects, it is necessary to cross-compare clinical trial results among individuals, treatment groups, and treatment-block interactions, as well as to results of trials involving drugs of similar effect. Additionally, administrators who monitor regulatory compliance and decide how to procure and allocate R&D resources must track trial results in relation to evolving regulatory constraints on human clinical trials and requirements for drug approval.

- **Functional Genomics Research** – Molecular biologists need to identify homologous genes and protein structures that show consistent patterns of co-occurrence among known, sequenced genomes. These patterns need to be examined through all levels of the taxonomic hierarchy, in relation to gene location, and/or degree of protein-structure similarity.

- **DNA-Array Expression Analyses** – Researchers use expression patterns to identify the roles of individual and interacting proteins in physiologic processes. These patterns are identified by comparing expression data across treatment groups, genomes or higher-level taxonomic groupings, within structural or functional groupings at the protein or organism level, and in relation to time.

- **Biological Systems Modeling** – A variety of probabilistic modeling techniques are used to identify functional relationships among genes and predict the physiologic role of unclassified proteins. Models are generated by results of previous experiments and must be compared to the results of new trials at a variety of levels of biological organization to evaluate their predictive capabilities.

OLAP technology can be used to quantify, tabulate, and model data within a multi-dimension organizational structure precisely as bioinformatics data is organized for cross comparison. The powerful computational properties of OLAP applications can summarize data (using any specified operation) at all hierarchical levels, as well as driving probabilistic models based on actual data inputs. Research applications can be linked and categorized, so administrators use the actual results that researchers are analyzing to quickly track progress in relation to regulatory constraints. Frequently, the genomic data needs to be combined with information in other datasets, such as patient history, types and course of treatments, and relations to biochemical techniques. In such case, OLAP servers can provide good ways to visualize the relation between the gene expression data in combination with other available datasets. The OLAP server can also be used as a visualization tool for visualizing the results of mining, particularly, the association rule mining. With such capability, scientists can analyze the gene expression across tissue type, protein (function) class and through various stages of the disease life cycle. This requires assembly of data from multiple relational tables.

OLAP applications can build OLAP cubes from identified "star schema" and is able to slice/dice, drill-through and roll-up results for several hierarchical structures. At the heart of a star schema is a fact table which links to various dimension tables. In case of a gene expression database, gene expression levels and indications of up regulated/down regulated can be thought of as facts (see Figure 22). Various dimensions that can be linked to this fact forming a star are tissue type, protein function class, disease stage, treatment info, patient age, species info, technique, chemical property data, chemical structure, *in vivo*, *in vitro* info etc.

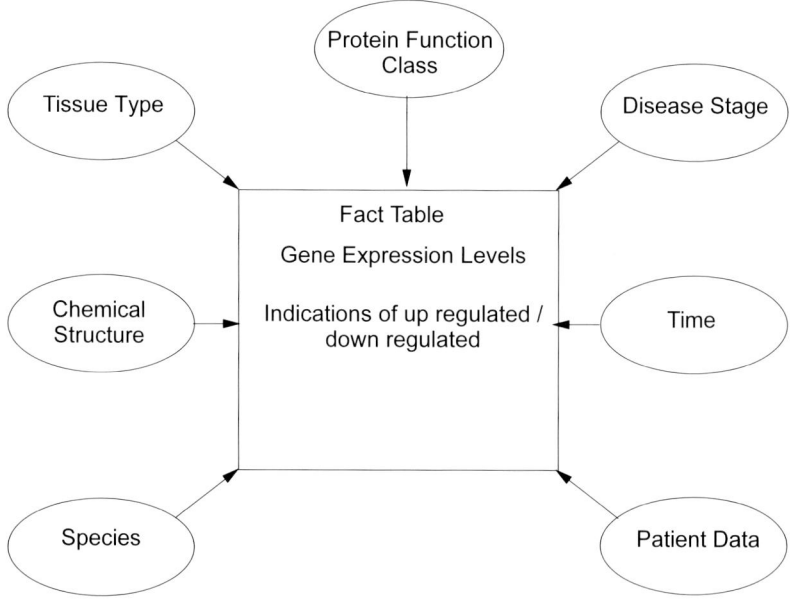

Figure 22. OLAP schema.

Data Mining

With the explosive growth of data in the genomic space, analysis of the data and turning it into information has become a daunting task. Simple statistics based techniques or simple numerical analysis is not good enough to discover useful and interesting information from the huge and ever growing datasets. Recently, data mining has been taking a significant role in this area. The main differentiation between statistics and mining is that statistics is hypotheses driven and mining is data driven. The basic difference in philosophy is central to their potential use and limitations for information retrieval and discovery in the genomic space.

There are several potential applications for mining in genomic space. Here, for simplicity two examples of DNA Sequences and Gene Expression Data are considered.

In DNA sequences, many statistical techniques, machine learning methods and minimum description length (MDL) principle have been used to find repeating patterns. However, most of these methods have been able to identify a proper subset of patterns that meet the specifications explicitly provided by the users. Mining on the other hand can be used for complete repeating patterns.

For gene expression data, there are many questions that SQL can answer or statistics can provide. However, as mentioned earlier, these are hypothesis/user driven activities and hence are limited. Mining on the other hand can answer questions like:

- What are the co-expression factors (contributory/inhibitory) for tissue of varying attributes (healthy/diseased).
- Identify some of the critical chemical properties that cause significant changes in gene expression levels (compound screening experiments).

Common data mining algorithms that have been used in biotech/genomic research:

1. Clustering (SOM, Neural networks)
2. PCA
3. Classification (neural, decision tree, KNN / k nearest neighbor)
4. Basic Statistics (regression, bivariate Analysis etc.)
5. Visualization tools (from profiles, e.g. SOMs / self-organizing maps, to networks, e.g. NAViGaTOR, ..)
6. Prediction Tools (neural)
7. Association rules
8. Time series Analysis
9. Similar Sequences
10. Long-Association Rules (Not in the public domain yet).
11. Clustering (demographic, HTP / high throughput profiles)
12. Integration – systems approach – from clinical and gene/protein expression, to CGH (comparative genome hybridization), SNP (single-nucleotide polymorphism), PPIs (protein-protein interactions).

It will be easy to demonstrate the standard common techniques like clustering and classification can be successfully used for genomic data analysis. One simple example of how distinguishing algorithm of existing IntelligentMiner can be useful is given below:

There are various genes like CDC04, CDC24 etc. for which gene expression data as function of time has been analyzed to find similar sequences (Figures 23 and 24). "Sequence name" is an IntelligentMiner term and does not represent sequence in genomic terms. Match fraction is the overall match of the gene expression data and number of similar sub-sequences is intuitive.

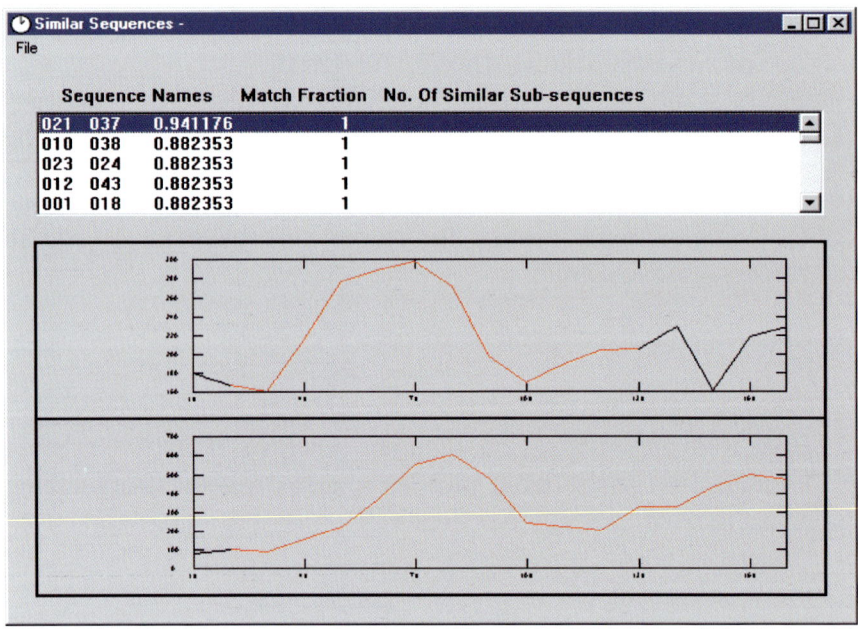

Figure 23. Similar sequences. CDC55 (index=21) and CDC5 (index=37) are shown to match with a match fraction of 0.94.

Informatics 127

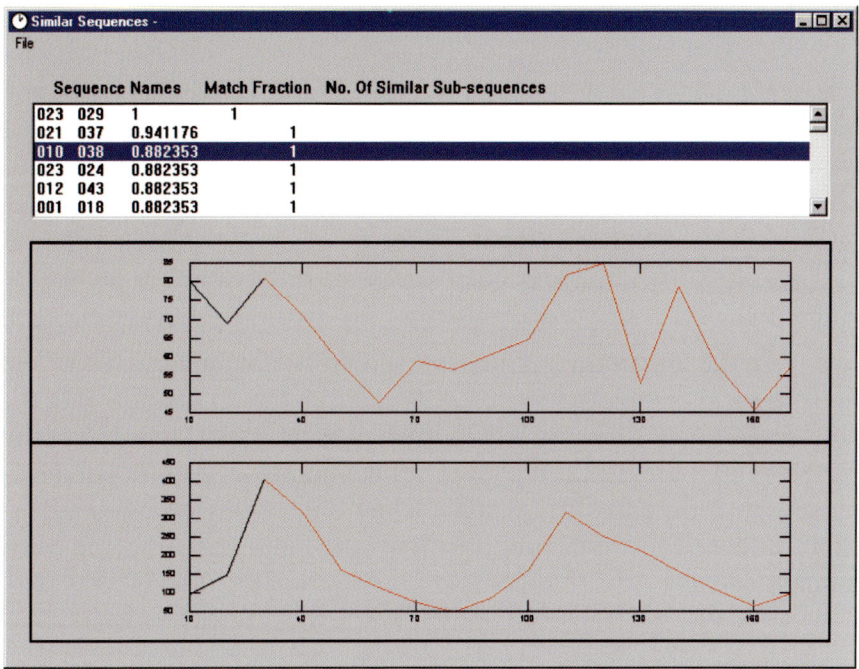

Figure 24. Similar sequences. CDC13 (index=10) and CDC17 (index =37) have a match of 0.8823 and this particular results are important because the scale of these two graphs is different and yet, IntelligentMiner has been able to find the similar sequence.

Multidimensional analysis is a good way to visualize and analyze gene expression data. The following questions that scientists would be asking like:

- Analyze gene expression level of a sequence (which maps to a protein or gene):

  across tissue type;
  across protein (function) class;
  through various stages of the disease lifecycle;
  types and course of treatments;
  in relation to biochemical technique; etc.

- Analyze gene expression as function of time and across certain tissue types, protein class and treatment types for exceptions.

Summary

Standards and new technologies are helping researchers become more efficient and effective in their quest to discover insights that are needed to develop targeted treatment solutions. Effective use of information technology has become the foundation for experimental biology. This trend is evident in genomics, proteomics, structural biology, and emerging areas associated with the study of metabolic regulation.

In the long term, researchers should be able to design and test drugs almost completely *in silico*. Predictive bio-simulation will enable researchers to work with virtual patients and "tune" specific variables in a biological model to reflect common genetic polymorphisms or differences in lifestyle. They will improve target validation, reduce lead times and attrition rates, and make testing with humans safer.

Integrative Computational Biology

Igor Jurisica

Addressing important clinical questions in cancer research will benefit from expanding computational biology. There is a great need to support systematic knowledge management and mining of the large amount of information to improve prevention, early diagnosis, cancer classification, prognostics and treatment planning, and to discover useful patterns.

Understanding normal and disease states of any organism requires integrated and systematic approach. We still lack understanding, and we are ramping up technologies to produce vast amounts of genomic and proteomic data. This provides both the opportunity and a challenge. No single database or algorithm will be successful at solving complex analytical problems. Thus, we need to integrate different tools and approaches, multiple single data type repositories, and repositories comprising diverse data types.

Knowledge management is concerned with the representation, organization, acquisition, creation, use and evolution of knowledge in its many forms. Effectively managing biological knowledge requires efficient representation schemas, flexible and scalable retrieval algorithms, robust and accurate analysis approaches and reasoning systems. We will discuss examples of how certain representation schemas support efficient retrieval and analysis, how the annotation and system integration can be supported using shareable and reusable ontologies, and how to manage tacit human knowledge.

Managing Biomedical Knowledge

Merely coping with the deluge of data is no longer an option; their systematic analysis is a necessity in the biomedical research. Computational biology is concerned with developing and using techniques from computer science, informatics, mathematics, and statistics to solve biological problems. Analyzing biomedical data requires robust approaches that deal with (ultra) high dimensionality, multimodal and rapidly evolving representations, missing information, ambiguity and uncertainty, noise, and incompleteness of domain theories. To diminish these problems, systematic knowledge management must be used, which includes the following main steps:

1. **Acquisition and preprocessing.** The system must acquire multiple types of data, including numeric, symbolic and image. Each of the data types conforms to standards (e.g. MIAME, PSI, etc.) and requires different preprocessing and normalization to enhance signal–to–noise ratio.

2. **Representation and organization.** To enable optimal (i.e. effective and efficient) use of acquired data, it must be represented and organized using database and knowledge base technologies. There is no one optimal representation though. Requirements change depending on the use of data. Sometimes flexibility and changing representation schema is of primary importance, other times, fast access is paramount.

3. **Integration and annotation.** Strong value of systems biology approach is to integrate multiple types of data, including gene and protein expression, chromosomal aberations, single nucleotide polymorphism, etc. To further enable interpretation of results, additional biological annotation databases can be used to furhter annotate the data, such as GeneOntology, protein interaction and pathway data, etc.

4. **Analysis.** Since there are multiple types of data, diverse algorithms must be used to analyze it.

5. **Visualization.** Complex data and results of analyses require intuitive visualization to aid knowledge discovery, hypothesis generation, and interpretation of results.

6. **Interpretation.** The final step in knowledge management is to interpret the results from analyses.

Advancing computational tools alone can improve each of these steps; however, this is not sufficient to impact computational biology and related biomedical fields. Many theoretically excellent approaches are inadequate for the high-throughput (HTP) biological domains, because of the scale or complexity of the problem, or because of the unrealistic assumptions on which they are based.

Acquisition and Preprocessing

The main challenges include diversity of data types, high dimensionality, missing information, incompatibility of individual platforms, large number of false positive and false negative rate of some experimental techniques, outliers due to lower technical consistency or due to intrinsic but unknown biological differences. This first step of biomedical data handling is sometimes referred to as sensor informatics (Lehmann, Aach et al. 2006).

The steps usually involve: image processing (i.e. image feature extraction), quality control and correction (may involve eliminating some data), normalization within and across array. For example, for microarray experiments one has to first extract important features from scanned images, and then normalize the data prior to further analysis. One of the main tasks is to remove spatial variance in data (Quackenbush 2002; Neuvial, Hupe et al. 2006; Yuan and Irizarry 2006; Yu, Nguyen et al. 2007). There are many well-established normalization methods (Bilban, Buehler et al. 2002; Quackenbush 2002; Yang, Dudoit et al. 2002; Bolstad, Irizarry et al. 2003; Cheadle, Vawter et al. 2003; Irizarry, Hobbs et al. 2003; Smyth and Speed 2003; Motakis, Nason et al. 2006), and several platform-specific (Neuvial, Hupe et al. 2006; Rabbee and Speed 2006; Wang, He et al. 2006).

Representation, Organization, and Integration

Truly understanding biological systems requires the integration of data across multiple high-throughput (HTP) platforms (Al-Shahrour, Minguez et al. 2005), including gene expression, protein abundance and interaction, and mutation information. Integrating heterogeneous and distributed data in a flexible manner is a challenging task. The goal is to achieve integration of flexibility provided by XML and RDF (Resource Description Framework) with rigidity and formal structure of existing ontologies (Almeida, Chen et al. 2006) Further, in order to prepare the infrastructure for modeling, we will represent dynamic and contextual aspects of information, such as interaction- or tissue-dependent localization, transient interactions, etc. (Scott, Calafell et al. 2005; Scott, Perkins et al. 2005).

Since no single database or algorithm will be successful at solving complex analytical problems, we must use multi-integration strategy to enable effective analysis and interpretation of cancer profiles. First, we can integrate multiple single data type repositories and repositories comprising diverse data types. Second, we integrate different algorithms for data mining and reasoning using genomic and proteomic cancer profiles, images, networks and text.

This information will get integrated with and annotated by public databases, such as Unigene, Genbank, GeneOntology, Locuslink, IPI (International Protein Index), SwissProt, PubMed, human and model organism protein-protein interaction (PPI) data sets and gene/protein expression, CGH (Comparative Genomic Hybridization), and SNP (Single Nucleotide Polymorphism) data sets.

For example, there are several resources for protein-protein interactions (see Table 1).

Table 1. Summary of some useful protein-protein interaction databases. Additional database are available at various lists, including JCB (http://www.imb-jena.de/jcb/ppi/jcb_ppi_databases.html) and NAR (http://www3.oup.co.uk/nar/database/cap), and http://www.biopax.org/.

Database	Name	URL	Reference
BIND	Biomolecular interaction network	http://bind.ca	(Bader, Betel et al. 2003)
BIOGRID	A general repository for interaction datasets	http://www.thebiogrid.org	(Breitkreutz, Stark et al. 2003)
DIP	Curated database of interacting proteins	http://dip.doe-mbi.ucla.edu	(Xenarios, Rice et al. 2000)
HPRD	Human reference protein interaction database	http://www.hprd.org	(Peri, Navarro et al. 2004)
HPID	Human Protein Interaction Database	http://wilab.inha.ac.kr/hpid	(Han, Park et al. 2004)
I2D	Interologous Interaction Database	http://ophid.utoronto.ca/i2d	
INTACT	Molecular interaction database	http://www.ebi.ac.uk/intact	(Al-Shahrour, Diaz-Uriarte et al. 2004; Hermjakob, Montecchi-Palazzi et al. 2004; Al-Shahrour, Minguez et al. 2005; Al-Shahrour, Minguez et al. 2006)
MINT	Molecular interaction database	http://cbm.bio.uniroma2.it/mint	(Zanzoni, Montecchi-Palazzi et al. 2002)
MIPS	Mammalian Protein-Protein Interaction Database	http://mips.gsf.de/proj/ppi	(Mewes, Frishman et al. 2002)

(Continued)

Table 1. (Continued)

OPHID	Online Predicted Human Interaction database – comprises predicted, experimental, and high-throughput interactions.	http://ophid.utoronto.ca	(Brown and Jurisica 2005)
POINT	Predicted and curated protein interaction database	http://point./bioinformatics.tw	(Huang, Tien et al. 2004)
STRING	Known and predicted protein interactions and associations	http://string.embl.de	(von Mering, Huynen et al. 2003)

Analysis, Visualization and Interpretation

Deriving useful knowledge from these data necessitates the creation of novel methods to store, analyze and visualize this information. Diverse statistical, machine learning and data mining approaches analyze each of the areas separately. The challenge is to develop innovative approaches that efficiently and effectively integrate and subsequently mine, visualize and interpret these various levels of information in a systematic and integrated fashion. Such strategies are necessary to model the biological questions posed by the complex phenotypes typically found in human disease such as cancer. The integration of data from multiple HTP methods is critical to understanding the molecular basis of normal organism function and disease.

Knowledge discovery is the process of extracting novel, useful, understandable and usable information from large data sets. In HTP biological domains, the first challenge is to deal with noise and high dimensionality. Often, dimensionality reduction using feature selection, principal component analysis or neural networks (Hinton 2000; Hinton and Salakhutdinov 2006) is an essential first step. However, care must be taken not to eliminate signal, albeit small, which may be essential when combined with other existing features. Thus, it is safer to start with feature-reduction rather then feature-selection approach. Since the goal is high degree of generality, we must use cross-validation on multiple, completely separate datasets for training and testing. Correctness can be determined by standard measures of cluster separation, such as Dunn's index, and by available biomedical annotation. However, annotations are frequently incorrect, and thus the combination of computational measures of outliers, and cluster separation and homogeneity needs to be considered together with annotation information. Some frequently used approaches include:

1. Association mining to derive novel, useful association rules describing data (Becquet, Blachon et al. 2002; Oyama, Kitano et al. 2002; Carmona-Saez, Chagoyen et al. 2006; Kotlyar and Jurisica 2006);
2. Self-organizing maps to support clustering of high-dimensional data and its visualization (Tamayo, Slonim et al. 1999; Toronen, Kolehmainen et al. 1999; Nikkila, Toronen et al. 2002; Sultan, Wigle et al. 2002; Brameier and Wiuf 2006) (see Figure 25);
3. Case-based reasoning to apply discovered knowledge and cancer signatures from association mining and self-organizing maps during interactive decision support (Macura, Macura et al. 1994; Ong, Shepherd et al. 1997; Jurisica, Mylopoulos et al. 1998; Bilska-Wolak and Floyd 2002; Jurisica and Glasgow 2004; Pantazi, Arocha et al. 2004; Rossille, Laurent et al. 2005).

Many other machine learning algorithms have been used in cancer informatics (Cruz and Wishart 2006).

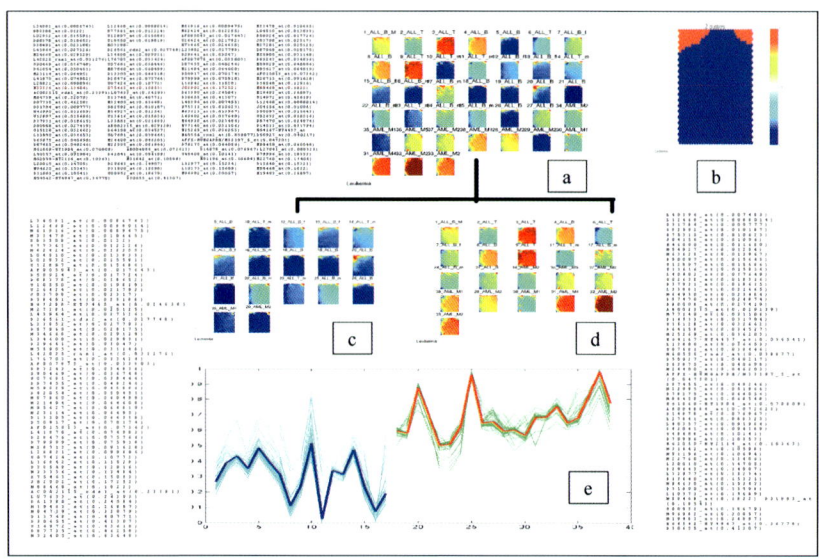

Figure 25. A typical node of BTSVQ algorithm: **(a)** (left) Quantized gene set, computed with SOM for all samples. (centre) Representation of gene expression of 38 samples for genes selected by vector quantization. **(b)** Genes selected by SOM are clustered by minimizing within cluster distance and maximizing intra cluster distance (Davuos Boulin distance measure). **(c)** (centre) Child one of the root node generated by partitive k-means algorithm, with k = 2. The visual representation of SOM component planes show that genes with lower levels of expression were separated from that with relatively high expression values by the partitive k-means algorithm. (left) Genes selected by vector quantization (using SOM) for the child one generated by partitive k-means algorithm. **(d)** Component planes and genes for child two. **(e)** Plot of genes selected by BTSVQ algorithm for a node.

In addition, feature selection algorithms can be used to reduce dimensionality. Support vector machines can be used to classify complex data into non-linear groups (Furey, Cristianini et al. 2000; Niijima and Kuhara 2005; Spinosa and Carvalho 2005; Pirooznia and Deng 2006). Decision trees are often use to support decision making process (Listgarten, Damaraju et al. 2004; Chen, Yu et al. 2007). Simple tools such as statistical correlation and clustering can also present useful trends in data (see Figure 26), but a more comprehensive array of algorithms will facilitate integrative anlaysis, e.g. (Simon, Lam et al. 2007).

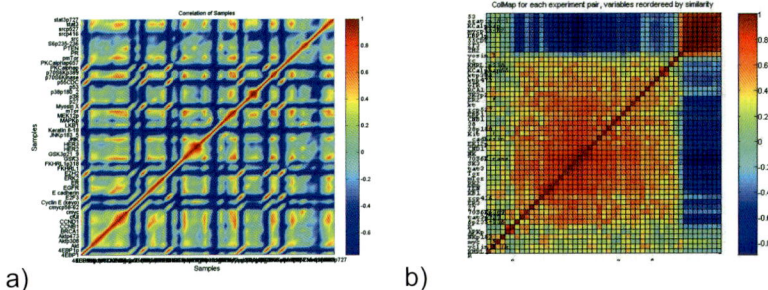

Figure 26. Pseudo-color correlation matrix clustering. **a)** Shows the original correlation data on target proteins. Since the targets were selected based on previous analysis and knowledge of involved pathways, targets nicely show the squares around the diagonal (it is a symmetric matrix, high positive correlation is dark red; negative correlation is blue). Importantly, there is a strong crosstalk among specific groups of proteins (rectangles off the diagonal). **b)** To systematically enable the interpretation of such results, the correlation matrix can be clustered to identify protein groups and inter group relationships.

These knowledge discovery and reasoning approaches can be combined with graph theory algorithms to derive structure–function relationship of protein interaction networks (Aittokallio, Kurki et al. 2003; King, Przulj et al. 2004; Przulj, Corneil et al. 2004; Przulj, Wigle et al. 2004). Text mining can be used for automated biomedical literature analysis and information extraction, such as ontology and taxonomy generation (Hirschman, Yeh et al. 2005), discovering relationships (Palakal, Stephens et al. 2002; Majoros, Subramanian et al. 2003; Liu, Navathe et al. 2005; Topinka and Shyu 2006), literature-based interaction and pathway validation (Hoffmann and Valencia 2005; Otasek, Brown et al. 2006) and database curation (Yeh, Hirschman et al. 2003; Miotto, Tan et al. 2005).

It has been established that despite inherent noise present in PPI data sets, systematic analysis of resulting networks uncovers biologically relevant information, such as lethality (Jeong, Mason et al. 2001; Hahn and Kern 2005), functional organization (Gavin,

Bosche et al. 2002; Maslov and Sneppen 2002; Sen, Kloczkowski et al. 2006; Wuchty, Barabasi et al. 2006), hierarchical structure (Ravasz, Somera et al. 2002; Lu, Shi et al. 2006; Yu and Gerstein 2006), dynamic modularity (Han, Bertin et al. 2004; de Aguiar and Bar-Yam 2005) and network-building motifs (Milo, Shen-Orr et al. 2002; Przulj, Corneil et al. 2004; Przulj, Corneil et al. 2006). These results suggest that PPI networks have a strong structure-function relationship (Przulj, Wigle et al. 2004), which we propose to use to help interpret integrated cancer profile data. Many PPIs are transient. Thus, the interaction networks change in different tissue, under different stimuli, or can be modified due to evolution (Barrios-Rodiles, Brown et al. 2005; Doyle, Alderson et al. 2005; Stefancic and Zlatic 2005; Takeuchi 2005; Wuchty and Almaas 2005). Studying the dynamic behavior of these networks, their intricacies in different tissue and under different stimuli is the exciting, but exponentially more complex, task that we are focused on. Extending the local structure analysis (Gao, Han et al. 2005) also suggests that the complex networks have self-organization dynamics (Sneppen, Bak et al. 1995). Many stable complexes show strong co-expression of corresponding genes, whereas transient complexes lack this support (Jansen, Greenbaum et al. 2002). This contextual dynamics of PPI networks must be considered when linking interaction networks to phenotypes, but also when studying the networks' topology. It is feasible to envision that while the current overall PPI network is best modeled by geometric random graphs (Przulj, Corneil et al. 2004), a different model may be needed to represent a transient network that is a subgraph of the original network. Adding this to different biases of individual HTP methods, the simple intersection of results achieves high precision at the cost of low recall. Systematic graph theory analysis of dynamic changes in PPI networks, combined with gene/protein cancer profiles will enable us to perform integrated analysis of cancer (Wachi, Yoneda et al. 2005; Zheng, Wang et al. 2005; Achard, Salvador et al. 2006; Aggarwal, Guo et al. 2006; Jonsson and Bates 2006; Jonsson, Cavanna et al. 2006; Kato, Murata et al. 2006; Kislinger and Jurisica 2006; Li, Wen et al. 2006; Pant and Ghosh 2006). Implementing algorithms using heuristics fine-tuned for PPI networks will ensure scalability.

Although several tools exist for visualizing graphs (Breitkreutz, Stark et al. 2002; Gilna 2002; Shannon, Markiel et al. 2003; Adai, Date et al. 2004; Han and Byun 2004; Han, Ju et al. 2004; Iragne, Nikolski et al. 2005; Kobourov and Wampler 2005), there is a need for a lightweight, OpenGL system for scalable 2D and 3D visualization and analysis of PPIs, combined with genomic and proteomic profiles, pathways from KEGG (Kanehisa, Goto et al. 2002), annotation from GO (Ashburner, Ball et al. 2000), and a flexible XML query language.

We can combine graph theoretic analysis of protein interaction networks with GeneOntology to provide annotation and generate hypotheses (see Figure 27).

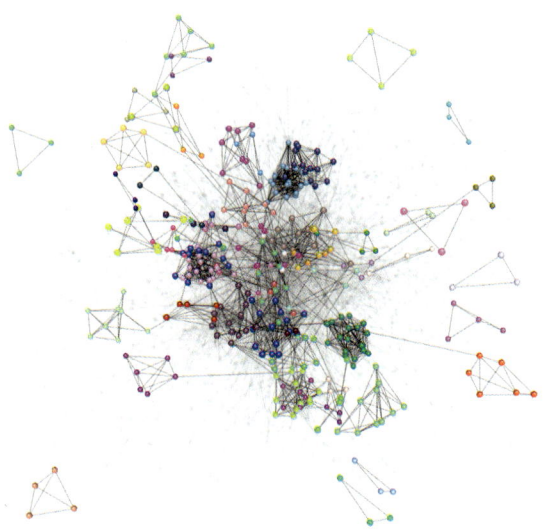

Figure 27. Visualization of protein complex data from (Collins, Kemmeren et al. 2007). Color represents cliques (highly interconnected protein complexes). Alpha-blending is used to suppress detail of the rest of the network. Visualized in 3D mode in NAViGaTor (http://ophid.utoronto.ca/navigator).

Combining interactions with microarray data enables to reduce noise in both data sets, and to predict new testable hypothesis (see Figure 28). Combining markers with protein interactions and known pathway enables us to annotate interaction data with direction, as shown in Figure 29.

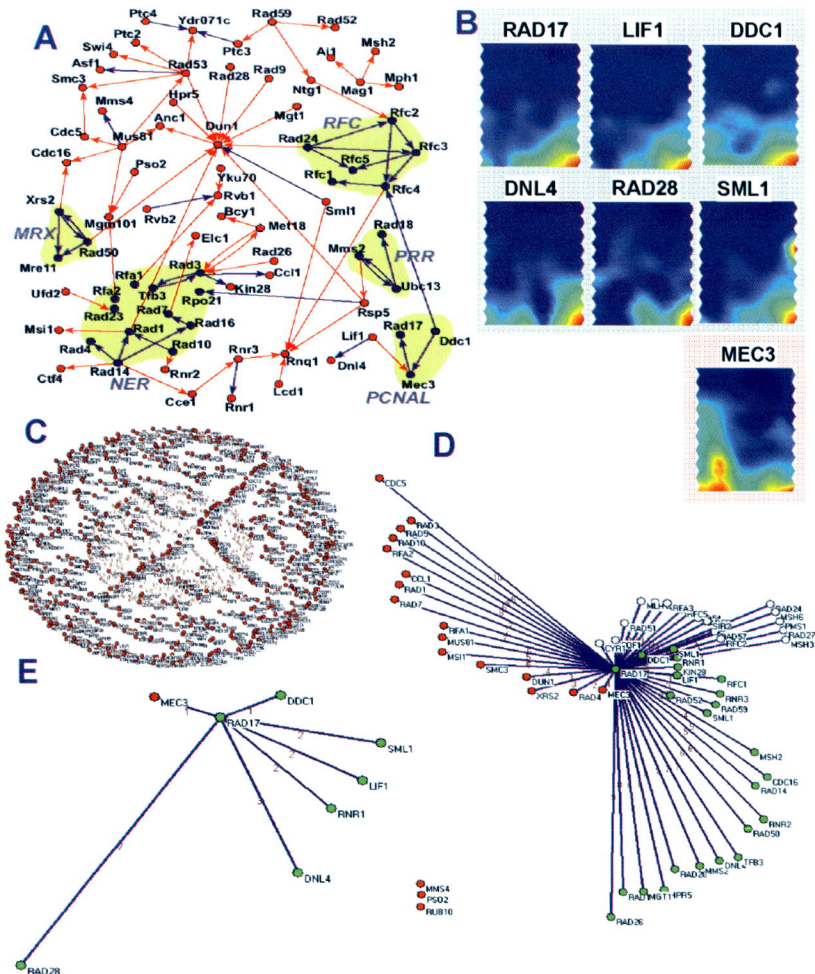

Figure 28. Integrated analysis of protein-protein interaction and microarray data. (A) Original DDR related PPI data from Figure 2 in (Ho, Gruhler et al. 2002). (B) Example of BTSVQ (Sultan, Wigle et al. 2002) analysis of yeast microarray data from (Hughes, Marton et al. 2000). (C) Graphical display of direct and indirect interactions of Rad17 with all 1,120 related proteins. (D) A weighted PPI graph that combines results from (A), (B), and (C) for Rad17. (E) A

hypothesis generated from integrated PPI and microarray data involving PCNA-like complex from (A).

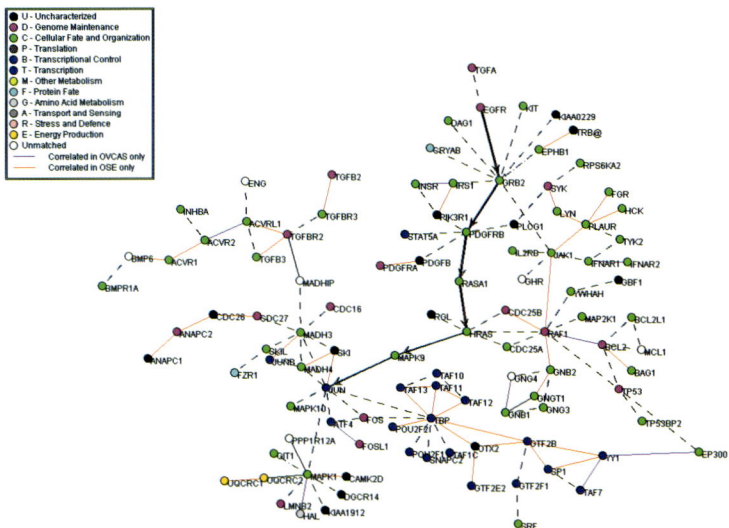

Figure 29. Integration of gene expression data with protein-protein interactions from OPHID (Brown and Jurisica 2005). The nodes in the network represent proteins; the color of the node represents annotated protein function when known (using GeneOntology). Lines connecting the nodes represent interactions between the two connected proteins. To emphasize interactions that are likely disrupted in cancer cells compared to non-malignant cells, in response to androgen, we use microarray data and co-expression between corresponding genes to "annotate" protein interactions. Black lines denote significantly correlated pairs in both groups, red lines denote correlation in cancer only, blue lines represent correlation in normal only, while dashed line represent no correlation. It clearly shows that there are full pathways and complexes that are only present in cancer samples (red lines). The highlighted (bold) line shows a known EGF pathway. Visualization done in NAViGaTor (http://ophid.utoronto.ca/navigator).

Medical Decision-Support Systems

The health care industry faces constant demands to improve quality and reduce cost. These demands are often addressed, within the field of medical informatics, by applying information technology to the delivery of health care services (Greenes and Shortliffe 1990). While many hospital information systems deal with simple tasks such as client billing, significant benefits can be realized by applying information systems to support medical decision-making.

Decision support in health care can be provided by knowledge-based systems. These systems make use of conventional information technology such as database management in conjunction with artificial intelligence techniques. However, the application of knowledge base technology to health care presents several challenges (Haux 1997). These challenges arise from the complexity of medical knowledge that is characterized by a large number of interdependent factors, the uncertainty of dependencies, and its constant evolution. It is imperative that medical decision support systems specifically address these challenges (Greenes and Shortliffe 1990).

Challenges of Knowledge Management

Knowledge-based systems go through the cycle of acquisition, representation and storage, analysis, and delivery of knowledge to the point of need.

Widespread application of information systems to health care faces two major difficulties: technical and cultural barriers (Brender, Nohr et al. 2000; Shortliffe and Sondik 2006; Pare and Trudel 2007). A major technical barrier is the lack of computing and communications standards. Cultural barriers arise from reengineering hospital operation, including interaction and information sharing among specialists, general practitioners, nurses, technicians and administrators. Therefore, successful implementation of IS requires a multidisciplinary approach. For example, physicians provide domain knowledge and define requirements as end users of such systems. Computer scientists architect the system infrastructure. Psychologists and linguists determine effective ways of communicating with end users.

Knowledge Representation

In several occasions, challenges in medical informatics have been discussed. It is agreed that knowledge representation is the core problem (Kahn and Huynh 1996; Haux 1997). These include knowledge representation, retrieval, visualization, analysis, and decision support.

Medical knowledge can be represented using various approaches. Selecting a particular formalism may require a tradeoff between information expressibility that the formalism supports, and scalability of the system that uses the formalism. In addition, effective knowledge representation formalism supports knowledge evolution and multiple contexts (Bellika, Hartvigsen et al. 2003; Verkoeijen, Rikers et al. 2004; Pantazi, Kushniruk et al. 2006).

Knowledge Retrieval and Delivery

The retrieval of medical knowledge is difficult because of its various forms, diversity of its location, and its potential to be contradictory (Bucci, Cagnoni et al. 1996; Chbeir, Amghar et al. 2001; Mao and Chu 2002; Dotsika 2003; Kagolovsky and Moehr 2003; Kagolovsky and Moehr 2003; Lehmann, Guld et al. 2004). In addition, there is a poor recall of journal-browsing doctors (Kellett, Hart et al. 1996). Results show that out of 75% of doctors who read a particular journal, only 48% correctly answered relevant questions. The second study reveals that out of 74% of doctors who read the journal, only 15% answered questions correctly. The recall rate can be improved by two approaches: by gaining more first-hand experience, and by using knowledge-based system to assist doctors during decision making process. Similarly, (Chiang, Hwang et al. 2006) showed that using SNOMED-CT to encode 242 concepts from five ophthalmology case presentations in a publicly-available clinical journal by three physicians results in 44% inter-coder.

Health care industry is lacking behind engineering and financial applications of decision-supports systems for several reasons. First, representing and managing medical information poses several technical problems, such as effectively representing patient records, and combining diverse health care information systems. Second, medical information is extremely sensitive and thus the privacy, security, and confidentiality issues must be addressed. Third, social and organizational aspects must be considered.

Future Directions

Despite the introduction of many powerful chemotherapeutic agents over the past two decades, most cancers retain devastating mortality rates. To significantly impact cancer research, novel therapeutic approaches for targeting metastatic disease and diagnostic markers reflective of changes associated with disease onset that can detect early stage disease must be discovered. Better drugs must be rationally designed, and current drugs made more efficacious either by re-engineering or by information-based combination therapy. To tackle these complex biological problems and impact HTP biology requires integrative computational biology, i.e. considering multiple data types, developing and applying diverse algorithms for heterogeneous data analysis and visualization. Improved analysis and reasoning algorithms will in turn advance disease diagnosis by finding better markers, and improve patient management by supporting information-based medicine.

Less than 50% of diagnosed cancers are cured using current treatment modalities. Many common cancers can already be fractionated into therapeutic subsets with unique prognostic outcomes based on characteristic molecular phenotypes. Combining molecular profiling and computational analysis will enable personalized medicine, where treatment strategies are individually tailored based on combinations of SNPs, gene expression and protein expression levels in biological samples. Integrating genomic and proteomic cancer profiles with PPIs enables: 1) objective target selection for validation, 2) implication of novel targets from the network that were not present in the original screen, 3) multi-target or weak target selection.

Although there is a substantial progress in the field by both improving experimental methods (i.e. increasing sensitivity and coverage of profiling platforms) and improving computational methods for handling resulting data, several challenges still remain. One of the main problems is that the techniques and tools can only work with the data measured. Thus, if the experiment was not planned properly, some questions will not be answered, or even worse, may be incorrectly inferred. This relates to for example heterogeneity of samples (e.g. non-standard processing of tissue, different tissue types, varying tumor content, etc.) (Blackhall, Pintilie et al. 2004), variation between platforms and quality control issues (Shi, Reid et al. 2006), improper experiment design (e.g. low power, insufficient replicates), and incorrect use of statistical and other analysis tools (Dupuy and Simon 2007).

To diminish some of these challenges, we need to intertwine experiment design–analysis–interpretation loop. This will not only improve data quality, but will also enable rational and unbiased hypothesis generation based on results from integrative analyses.

Part IV – Future Steps and Challenges

Igor Jurisica and Dennis A. Wigle

Medical information science involves using system-analytic tools to develop algorithms for management, process control, decision-making, and scientific analysis of medical knowledge. Medical informatics comprises the theoretical and practical aspects of information processing and communication, based on knowledge and experience derived from processes in medicine and health care. This is achieved by developing and assessing methods and systems to support the acquisition, processing and interpretation of patient data with the help of knowledge that is obtained from basic and clinical research.

Medical practitioners have been treating patients by integrating knowledge and best practices, personal experience and clinical observation since the days of Hippocrates. However, exponential increases in the body of knowledge applicable to patient care have resulted in ever increasing niche specialization in large, academic, tertiary care medical centers. In modern cancer care, the days of the "generalist" are long gone. While further specialization in areas of expertise is possible, the need for computational approaches to knowledge discovery, information management, and decision support continue to increase. While expertise is essential, advancing available "tools" and methods has the potential to revolutionize many aspects of healthcare delivery. Recently, we have witnessed an accelerated understanding of complex diseases at the molecular level. Cancer informatics provides both a methodology and tools to handle such information on a patient-centered level. Although many challenges remain ahead, this progress toward information-based medicine has the potential to increase healthcare quality and enable innovative approaches in a true personalized manner. Key challenges include the development of comprehensive electronic patient records and biobank repositories, data integration and sharing, and seamless integration and translation of research into clinical care.

Advancing from histopathologically based disease classifications to true molecular staging based on genomic and proteomic profiles will require ongoing development of novel computational tools for clinical correlation. Many of the tools developed to date represent a major step forward in cancer informatics, but further development will be required to enable routine clinical application. Given the molecular heterogeneity of cancer, this is an obvious area to integrate and analyze diverse data sets for their ability to provide additional information. These integrated analyses of multidimensional data will reveal markers that enhance existing clinical approaches to diagnosis, prognosis and treatment planning in cancer. The development of cancer profiles could potentially lead to new cancer treatments as well as techniques for early diagnosis. The long-term goal of these collective strategies is information-based individualized patient care. There are already low-throughput examples of genotyping for genetic markers (e.g. cystic fibrosis) and profiling for disease markers (e.g. prostate-specific antigen).

Measuring the genomic expression profile in cell cultures and accumulating a set of characteristic profiles as a background information base can assess the effect of known toxic compounds. Patient progress can be assessed by detailed measurements of thousands of molecular indicators from bodily fluids, biopsies, such as RNA expression, protein expression, protein modification, or concentration of metabolites. However, the current medical practice is primarily reactive – frequently, we only treat disease after symptoms appear, which for cancer usually means an advanced stage with dismal prognosis and limited treatment options. Even when the treatments are available, we may not deliver them optimally for an individual patient. The FDA's Center for Drug Evaluation and Research estimates that approximately 2 million of the 2.8 billion prescriptions filled annually in the United States will result in adverse drug reactions, leading to about 100,000 deaths per year. To diminish these problems, we have to further:

1. Accelerate the molecular understanding of cancer by systems biology approaches to investigate the underlying basis of disease.
2. Extend and apply cancer informatics to support the acquisition, integration, analysis, visualization, interpretation, and dissemination of integrated molecular and clinical data for decision support.

One of the current goals from amassing large databases of protein structural information is the ability to compute reliable structural predictions of proteins based on amino acid sequence. The attainment of this goal would greatly facilitate the design of synthetic organic compounds in medicinal chemistry and dramatically accelerate the pace of rational drug design. Speeding up this process of lead target to drug candidate is a critical step in translating the volume of high-throughput data being generated in disease models to clinical utility.

True understanding of biological systems will require the integration of data across multiple high-throughput platforms. Our ability to derive true knowledge from the current data being generated on SNPs, gene expression, protein abundance and interaction, and mutational information will necessitate the creation of novel methods to store, analyze, and visualize this information. The advent of genomic and proteomic technologies has ushered forth the era of genomic medicine. The promise of these advances is true "personalized medicine" where treatment strategies can be individually tailored based on combinations of SNPs, gene expression, and protein expression levels in biological samples. Translating these advances to the improvement of objective outcomes such as prolonged survival and increased quality of life is eagerly awaited by patients with cancer and their healthcare providers.

Glossary

Association learning
Techniques that find conjunctive implication rules (associations) that satisfy given criteria. The conventional association algorithms are sound and complete methods for finding all associations that satisfy criteria for minimum support (at least this fraction of the instances must satisfy both sides) and minimum confidence (at least this fraction of instances satisfying the left hand side, or antecedent, must satisfy the right hand side, or consequent).

Attribute
A quantity describing an instance (feature). An attribute has a domain, which denotes the values that can be taken by an attribute – the attribute's type. The following domain types are common: Nominal (categorical). No relation holds between different values. For example: last name, color. Ordinal. There is a known ordering to the values. Continuous. Subset of real numbers. Integers are usually treated as continuous in machine learning work. Discrete. There is a finite set of values.

Bioinformatics
The application of computational techniques to biology, in particular molecular biology.

The Cancer Biomedical Information Grid - Cancer Bioinformatics Infrastructure Objects (caBIG - caBIO)
The caBIO model and architecture is the primary programmatic interface to caCORE. The heart of caBIO is its domain objects, each of which represents an entity found in biomedical research. These domain objects are related to each other, and examining these relationships can bring to the surface biomedical knowledge that was previously buried in the primary data sources (http://ncicb.nci.nih.gov/core/caBIO)

The Cancer Biomedical Information Grid - Cancer Genome Anatomy Project (caBIG - cGAP)
The information in the Mitelman Database of Chromosome Aberrations in Cancer relates chromosomal aberrations to tumor characteristics, based either on individual cases or associations. cGAP has developed five web search tools to help analyze the information within the Mitelman Database (http://cgap.nci.nih.gov/)

Cancer informatics
Cancer informatics provides both a methodology and practical information tools. Cancer informatics supports a patient-centric record with access to personalized protocols and clinical guidelines supported by and being part of a continuously updated clinical trials system.

Case-Based Reasoning
A reasoning paradigm that solves new problems by reusing solutions from past similar problems.

Classifier
A system that performs automatic classification.

Clinical trials and longitudinal studies
A clinical trial is a scientific study to determine the safety and effectiveness of a treatment or intervention. A longitudinal study is a study in which the same group of individuals is monitored at intervals over a period of time.

Clustering
The process of grouping data points into clusters, i.e. a set of data points that are grouped by their proximity in a metric space.

Collision induced dissociation (CID)
The fragmentation of ions by collision with inert gas molecules

Computational biology
The development of computational tools and efficient algorithms for biology to support data base management, search, analysis and knowledge discovery, mathematical modeling, computer simulation, etc.

Cytogenetics
The study of chromosomes and chromosome abnormalities.

Data mining
The process of finding useful patterns in data.

Data model
A mathematical formalism comprising a notation for describing data and a set of operations used to manipulate that data.

The database of single nucleotide polymorphisms (dbSNP). The database of single nucleotide polymorphisms (http://www.ncbi.nlm.nih.gov/SNP/).

Glossary

Detection depth
The number of proteins detected in a proteomics experiment

Differential in-gel electrophoresis (DIGE)
Quantitative proteomics by 2-DE based on selective fluorescence labeling of proteins. Protein extracts are labeled with two different fluorophores, combined and separated by 2-DE. Differential excitation of the two fluorophores allows for relative quantification.

DNA
The sequencing, identification of genes, analysis of genetic variation and mutation analysis. Technologies include DNA sequencing, phylogenetics, haplotyping and SNP identification.

Edman sequencing
A methodology developed by Pehr Edman, also known as Edman degradation. It is the selective, step-by-step removal of the N-terminal amino acid of a protein after selective modification with phenyl isothiocyanate. This step will be repeated to identify stretches of the amino acid sequence.

Electrospray ionization (ESI)
A mild ionization form used for biomolecular MS. A liquid containing the analyte of interest is pumped through a narrow column. A high voltage is applied directly to the solvent and a fine aerosol of droplets is sprayed from the end of the column.

Epidemiology and population studies
Epidemiological and population research studies investigate the incidence, distribution, and control of disease in a population.

Evidence-based medicine (EBM)
The integration of best research evidence with clinical expertise and patient values.

Functional genomics
The exploration and analysis of gene function. Technologies include microarray, ChIP, and network analysis.

The Gene Ontology (GO)
The GO is a controlled vocabulary of biological processes, functions and localizations. (http://www.geneontology.org/).

Information-based medicine
Integration of healthcare, life sciences, and information technology with the goal to deliver relevant information to researchers and clinicians in real time, support acquisition, integration, analysis, visualization and interpretation of complex data.

Isotope coded affinity tags (ICAT)
A chemical isotope label for quantitative proteomics. The reagent specifically reacts with cysteine SH-side chains.

Isotope coded protein label (ICPL)
A chemical isotope label for quantitative proteomics. The reagent specifically reacts with lysine NH2-side chains.

Kyoto Encyclopedia of Genes and Genomes (KEGG) KEGG is a suite of databases and associated software covering the information about the genes, proteins, chemical compounds and reactions, and molecular pathways (http://www.genome.ad.jp/kegg/kegg2.html).

Knowledge discovery
A (nontrivial) process of identifying valid, novel, potentially useful, and ultimately understandable patterns in data.

Knowledge management
The representation, organization, acquisition, creation, use and evolution of knowledge in its many forms.

Laser capture microdissection (LCM)
A methodology for the isolation of selected cells from solid tissue with a low power laser beam.

Mascot
Commercially available search algorithm from Matrix Science.

Mass spectrometer (MS)
An instrument used to measure the mass-to-charge ratio (m/z) of ions in the gas phase.

Liquid chromatography mass spectrometry (LC-MS)
The coupling of liquid chromatography with mass spectrometry

Matrix-assisted laser desorption/ionization (MALDI-TOF-MS)
A mild ionization form used for biomolecular MS. An analyte is mixed with a matrix molecule and crystallized on top of a sample target plate. A pulsing laser is used to ionize both the analyte of interest and the matrix molecule.

Surface-enhanced laser desorption ionization time-of-flight mass spectrometry (SELDI-TOF-MS)
A variation of MALDI-TOF-MS. The sample target plate is coated with different chromatography resins (e.g. ion exchange, reverse phase etc.) to minimize the

sample complexity. The systems were commercially introduced by Ciphergen Biosystems.

Tandem mass spectrometry (MS/MS)
First the mass of an ion is determined by MS, then individual ions are isolated and fragmented by collision induced dissociation and the m/z of each fragmentation peak is determined. The result is a tandem mass spectrum (MS/MS)

Medical informatics
Medical information science involves using system-analytic tools to develop algorithms for management, process control, decision-making, and scientific analysis of medical knowledge. Medical informatics comprises the theoretical and practical aspects of information processing and communication, based on knowledge and experience derived from processes in medicine and health care.

Metabolomics
Large-scale detection of small molecular metabolites.

Model
A characterization of relationships between input and output variables.

Multidimensional protein identification technology (MudPIT)
The on-line separation of peptides by two-dimensional chromatography followed by mass spectrometry

mzXML
An open, generic XML representation of mass spectrometry data (http://sashimi.sourceforge.net/software.html).

OMSSA
The Open Mass Spectrometry Search Algorithm (OMSSA) is available from the NCBI.

Ontology Web language (OWL) A semantic markup language for creatinhg and sharing ontologies on the World Wide Web. OWL has been developed as an extension of RDF (Resource Description Framework). It is derived from the DAML+OIL Web Ontology language and is a collaborative development endorsed by the W3C (http://www.w3.org/TR/2004/REC-owl-ref-20040210/).

Outlier
An example pattern that is not representative of the majority of observed data.

Peptide mass fingerprint
Proteins are digested with sequence specific enzymes. The collection of resulting peptide ions can be used as an identifier of the unknown protein.

Proteome
The complete set of proteins expressed by an organism

Proteomics
Large-scale identification, characterization and quantification of proteins involved in a particular pathway, organelle, cell, tissue, organ or organism that can be studied in concert to provide accurate and comprehensive data about that system. Technologies include protein interaction models, high-throughput protein analysis and modeling.

Expression proteomics
Detection of every protein expressed in a given biological system.

Functional proteomics
Global detection of protein-protein interactions.

Proteomics of posttranslational modifications
Detection of posttranslational protein modifications by mass spectrometry. The ultimate goal is to detect every modification, changes in its relative abundance and its exact localization on each protein.

Structural proteomics
The determination of the exact three-dimensional structure of every protein.

Proteomics investigation strategy for Mammals (PRISM)
An analysis strategy for mammalian proteins by LC-MS. Mammalian tissue is fractionated into discrete organelle fractions. Proteins are analyzed by MudPIT profiling, generated spectra are statistically validated and high confidence proteins are automatically mapped against the Gene Ontology annotation scheme.

Resource Description Framework (RDF)
A language for representing information about resources in the World Wide Web. It is intended for representation of metadata about Web resources, such as the title or author of a web page (http://www.w3.org/TR/2004/REC-rdf-primer-20040210/#intro).

Reactome
Reactome covers biological pathways ranging from the basic processes of metabolism to high-level processes such as hormonal signaling, with a special emphasis on human data (http://www.reactome.org/).

Systems Biology Markup Language (SBML) A format designed to enable the exchange of biochemical network models between different software packages (http://www.sbml.org/docs/).

Glossary 157

SEQUEST
Commercially available search algorithm from Thermo Finnigan. The original code was developed by Yates and colleagues.

Silica-bead perfusion technique
Selective isolation of plasma membranes by coating with colloidal cationic silica beads

Snap-frozen
Quick freeze of tissue/cells by placing them immediately in liquid nitrogen or isopentane-dry ice.

Simple Object Access Protocol (SOAP)
An XML based protocol for encoding of application defined data types from a number of different languages, and remote procedure calls and responses. SOAP enables different programs running on different computers to communicate with one another.

Stable Isotope Labeling with Amino acids in Cell culture (SILAC)
An isotope labeling strategy for proteins in cell culture for quantitative proteomics. The cell culture growth medium is supplemented with an essential nutrient (e.g. an amino acid) in either the light or heavy isotopic version.

Systems biology
Scientific approach that uses quantitative systems-level experimentation to systematically identify and characterize molecules and molecular interactions that define cellular pathways, tissues, organs and organisms, and applies computational biology tools to integrate, analyze, visualize, and model experimental data.

Therapeutics
The treatment of cancer by remedial agents or methods.

Tissue Micro Array (TMA)
A platform for high throughput analysis and examination of a large number of tumor cases on a single histology slide and having been subjected to a specific stain. In TMA, small (6-15 mm diameter) cores of formalin fixed and paraffin embedded tissue are arrayed into a single paraffin block.

Transcriptomics
Global detection of mRNA

Two-dimensional gel electrophoresis (2-DE)
Separation of proteins by gel electrophoresis in two dimensions. In the first dimension proteins are separated by isoelectric point. In the second dimension proteins are separated based on molecular mass.

Xenograft
Growth of cells/tissues in organism/animal of different species, e.g. human tumor cells in immune deficient mice.

X!Tandem
An open-source proteomics search algorithm available from the Global Proteome Machine Organization (http://www.thegpm.org)

Extensible Markup Language (XML)
A flexible and adaptable information identification format (www.xml.org).

References

Achard, S., R. Salvador, et al. (2006). "A resilient, low-frequency, small-world human brain functional network with highly connected association cortical hubs." *J Neurosci* **26**(1): 63-72.

Adai, A. T., S. V. Date, et al. (2004). "LGL: creating a map of protein function with an algorithm for visualizing very large biological networks." *J Mol Biol* **340**(1): 179-90.

Adams, J. and S. Cory (1991). "Transgenic models of tumor development." *Science* **254**(5035): 1161.

Aebersold, R. and M. Mann (2003). "Mass spectrometry-based proteomics." *Nature* **422**(6928): 198-207.

Aggarwal, A., D. L. Guo, et al. (2006). "Topological and functional discovery in a gene coexpression meta-network of gastric cancer." *Cancer Res* **66**(1): 232-41.

Aittokallio, T., M. Kurki, et al. (2003). "Computational strategies for analyzing data in gene expression microarray experiments." *J Bioinform Comput Biol* **1**(3): 541-86.

Al-Shahrour, F., R. Diaz-Uriarte, et al. (2004). "FatiGO: a web tool for finding significant associations of Gene Ontology terms with groups of genes." *Bioinformatics* **20**(4): 578-80.

Al-Shahrour, F., P. Minguez, et al. (2006). "BABELOMICS: a systems biology perspective in the functional annotation of genome-scale experiments." *Nucleic Acids Res* **34**(Web Server issue): W472-6.

Al-Shahrour, F., P. Minguez, et al. (2005). "BABELOMICS: a suite of web tools for functional annotation and analysis of groups of genes in high-throughput experiments." *Nucleic Acids Res* **33**(Web Server issue): W460-4.

Alizadeh, A., M. Eisen, et al. (2000). "Distinct types of diffuse large B-cell lymphoma identified by gene expression profiling." *Nature* **403**: 503-511.

Allgayer, H. (2003). "Molecular staging of cancer: concepts of today, therapies of tomorrow." *J Surg Oncol* **82**(4): 217-23.

Almeida, J. S., C. Chen, et al. (2006). "Data integration gets 'Sloppy'." *Nat Biotechnol* **24**(9): 1070-1.

An, Z., X. Wang, et al. (1996). "A clinical nude mouse metastatic model for highly malignant human pancreatic cancer." *Anticancer Res* **16**(2): 627-31.

Anderson, L. and L. Priest (1980). "Reduction in the transplacental carcinogenic effect of methylcholanthrene in mice by prior treatment with beta-naphthoflavone." *Res Commun Chem Pathol Pharmacol* **30**(3): 431-46.

Anderson, N. L., M. Polanski, et al. (2004). "The human plasma proteome: a non-redundant list developed by combination of four separate sources." *Mol Cell Proteomics* **3**(4): 311-26.

Arteaga, C. L., F. Khuri, et al. (2002). "Overview of rationale and clinical trials with signal transduction inhibitors in lung cancer." *Semin Oncol* **29**(1 Suppl 4): 15-26.

Ashburner, M., C. A. Ball, et al. (2000). "Gene ontology: tool for the unification of biology. The Gene Ontology Consortium." *Nat Genet* **25**(1): 25-9.

Bader, G. D., D. Betel, et al. (2003). "BIND: the Biomolecular Interaction Network Database." *Nucleic Acids Res* **31**(1): 248-50.

Barrios-Rodiles, M., K. R. Brown, et al. (2005). "High-throughput mapping of a dynamic signaling network in mammalian cells." *Science* **307**(5715): 1621-5.

Becquet, C., S. Blachon, et al. (2002). "Strong-association-rule mining for large-scale gene-expression data analysis: a case study on human SAGE data." *Genome Biol* **3**(12): RESEARCH0067.

Beer, D., S. Kardia, et al. (2002). "of patients with lung adenocarcinoma." *NATURE MEDICINE* **8**(8): 816-824.

Beer, D. G., S. L. Kardia, et al. (2002). "Gene-expression profiles predict survival of patients with lung adenocarcinoma." *Nat Med* **8**(8): 816-24.

Belinsky, S. (1993). "The A/J mouse lung as a model for developing new chemointervention strategies." *Cancer Research* **53**(2): 410-416.

Bellika, J. G., G. Hartvigsen, et al. (2003). "Using discharge letters and context representation in information retrieval of medical literature." *Stud Health Technol Inform* **95**: 373-8.

Bhattacharjee, A., W. Richards, et al. (2001). "Classification of Human Lung Carcinomas by mRNA Expression Profiling Reveals Distinct Adenocarcinoma Subclasses." *Proceedings of the National Academy of Sciences of the United States of America* **98**(24): 13790-13795.

Bhattacharjee, A., W. G. Richards, et al. (2001). "Classification of human lung carcinomas by mRNA expression profiling reveals distinct adenocarcinoma subclasses." *Proc Natl Acad Sci U S A* **98**(24): 13790-5.

Bilban, M., L. K. Buehler, et al. (2002). "Normalizing DNA microarray data." *Curr Issues Mol Biol* **4**(2): 57-64.

Bild, A., G. Yao, et al. (2006). "Oncogenic pathway signatures in human cancers as a guide to targeted therapies." *Nature* **439**(7074): 353-357.

Bild, A. H., G. Yao, et al. (2006). "Oncogenic pathway signatures in human cancers as a guide to targeted therapies." *Nature* **439**(7074): 353-7.

Bilska-Wolak, A. O. and C. E. Floyd, Jr. (2002). "Development and evaluation of a case-based reasoning classifier for prediction of breast biopsy outcome with BI-RADS lexicon." *Med Phys* **29**(9): 2090-100.

Blackhall, F., M. Pintilie, et al. (2004). "Stability and heterogeneity of expression profiles in lung cancer specimens harvested following surgical resection." *Neoplasia* **6**(6): 761-7.

Blackhall, F., M. Ranson, et al. (2006). "Where next for gefitinib in patients with lung cancer?" *Lancet Oncology* **7**(6): 499-507.

Blackhall, F. H., M. Pintilie, et al. (2004). "Stability and heterogeneity of expression profiles in lung cancer specimens harvested following surgical resection." *Neoplasia* **6**(6): 761-7.

Bolstad, B. M., R. A. Irizarry, et al. (2003). "A comparison of normalization methods for high density oligonucleotide array data based on variance and bias." *Bioinformatics* **19**(2): 185-93.

Boven, E. (1992). "Phase II preclinical drug screening in human tumor xenografts: a first European multicenter collaborative study." *Cancer Research* **52**(21): 5940-5947.

Brameier, M. and C. Wiuf (2006). "Co-clustering and visualization of gene expression data and gene ontology terms for Saccharomyces cerevisiae using self-organizing maps." *J Biomed Inform*.

Breitkreutz, B. J., C. Stark, et al. (2002). "Osprey: a network visualization system." *Genome Biol* **3**(12): PREPRINT0012.

Breitkreutz, B. J., C. Stark, et al. (2003). "The GRID: the General Repository for Interaction Datasets." *Genome Biol* **4**(3): R23.

Brender, J., C. Nohr, et al. (2000). "Research needs and priorities in health informatics." *Int J Med Inform* **58-59**: 257-89.

Brierley, M. M., K. L. Marchington, et al. (2006). "Identification of GAS-dependent interferon-sensitive target genes whose transcription is STAT2-dependent but ISGF3-independent." *Febs J* **273**(7): 1569-81.

Brinster, R., H. Chen, et al. (1985). "Factors Affecting the Efficiency of Introducing Foreign DNA into Mice by Microinjecting Eggs." *Proceedings of the National Academy of Sciences* **82**(13): 4438-4442.

Brown, K. R. and I. Jurisica (2005). "Online predicted human interaction database." *Bioinformatics* **21**(9): 2076-82.

Bucci, G., S. Cagnoni, et al. (1996). "Integrating content-based retrieval in a medical image reference database." *Comput Med Imaging Graph* **20**(4): 231-41.

Butler, T. (1975). "Quantitation of cell shedding into efferent blood of mammary adenocarcinoma." *Cancer Research* **35**(3): 512-516.

Caldwell, R. L. and R. M. Caprioli (2005). "Tissue profiling by mass spectrometry: a review of methodology and applications." *Mol Cell Proteomics* **4**(4): 394-401.

Callister, S. J., R. C. Barry, et al. (2006). "Normalization approaches for removing systematic biases associated with mass spectrometry and label-free proteomics." *J Proteome Res* **5**(2): 277-86.

Carmona-Saez, P., M. Chagoyen, et al. (2006). "Integrated analysis of gene expression by Association Rules Discovery." *BMC Bioinformatics* **7**: 54.

Chaurand, P., D. S. Cornett, et al. (2006). "Molecular imaging of thin mammalian tissue sections by mass spectrometry." *Curr Opin Biotechnol* **17**(4): 431-6.

Chaurand, P., S. A. Schwartz, et al. (2002). "Imaging mass spectrometry: a new tool to investigate the spatial organization of peptides and proteins in mammalian tissue sections." *Curr Opin Chem Biol* **6**(5): 676-81.

Chbeir, R., Y. Amghar, et al. (2001). "A four-dimensional approach to medical image retrieval." *Methods Inf Med* **40**(3): 178-83.

Cheadle, C., M. P. Vawter, et al. (2003). "Analysis of microarray data using Z score transformation." *J Mol Diagn* **5**(2): 73-81.

Chen, H. Y., S. L. Yu, et al. (2007). "A five-gene signature and clinical outcome in non-small-cell lung cancer." *N Engl J Med* **356**(1): 11-20.

Chiang, M. F., J. C. Hwang, et al. (2006). "Reliability of SNOMED-CT Coding by Three Physicians using Two Terminology Browsers." *AMIA Annu Symp Proc*: 131-5.

Collins, S. R., P. Kemmeren, et al. (2007). "Towards a comprehensive atlas of the physical interactome of Saccharomyces cerevisiae." *Mol Cell Proteomics*.

Corbett, T. (1975). "Tumor induction relationships in development of transplantable cancers of the colon in mice for chemotherapy assays, with a note on carcinogen structure." *Cancer Research* **35**(9): 2434-2439.

Corbett, T. (1984). "Induction and chemotherapeutic response of two transplantable ductal adenocarcinomas of the pancreas in C57BL/6 mice." *Cancer Research* **44**(2): 717-726.

Craig, R. and R. C. Beavis (2004). "TANDEM: matching proteins with tandem mass spectra." *Bioinformatics* **20**(9): 1466-7.

Cruz, J. A. and D. S. Wishart (2006). "Applications of Machine Learning in Cancer Prediction and Prognosis." *Cancer Informatics* **2**(2): 59-77.

Dancik, V., T. A. Addona, et al. (1999). "De novo peptide sequencing via tandem mass spectrometry." *J Comput Biol* **6**(3-4): 327-42.

Davies, B. (1993). "A synthetic matrix metalloproteinase inhibitor decreases tumor burden and prolongs survival of mice bearing human ovarian carcinoma xenografts [published erratum appears in Cancer Res 1993 Aug 1; 53 (15): 3652]." *Cancer Research* **53**(9): 2087-2091.

de Aguiar, M. A. and Y. Bar-Yam (2005). "Spectral analysis and the dynamic response of complex networks." *Phys Rev E Stat Nonlin Soft Matter Phys* **71**(1 Pt 2): 016106.

Denoix, P. (1944). "Tumor, node, and metastasis (TNM)." *Bull Inst Nat Hyg (Paris)* **1**: 1-69.

DeVita, V. and P. Schein (1973). "The use of drugs in combination for the treatment of cancer: rationale and results." *N Engl J Med* **288**(19): 998-1006.

Dinney, C., R. Fishbeck, et al. (1995). "Isolation and characterization of metastatic variants from human transitional cell carcinoma passaged by orthotopic implantation in athymic nude mice." *J Urol* **154**(4): 1532-8.

Domon, B. and R. Aebersold (2006). "Mass spectrometry and protein analysis." *Science* **312**(5771): 212-7.

Dotsika, F. (2003). "From data to knowledge in e-health applications: an integrated system for medical information modelling and retrieval." *Med Inform Internet Med* **28**(4): 231-51.

Doyle, J. C., D. L. Alderson, et al. (2005). "The "robust yet fragile" nature of the Internet." *Proc Natl Acad Sci U S A* **102**(41): 14497-502.

Duperron, C. and A. Castonguay (1997). "Chemopreventive efficacies of aspirin and sulindac against lung tumorigenesis in A/J mice." *Carcinogenesis* **18**(5): 1001-1006.

Dupuy, A. and R. M. Simon (2007). "Critical review of published microarray studies for cancer outcome and guidelines on statistical analysis and reporting." *J Natl Cancer Inst* **99**(2): 147-57.

Durr, E., J. Yu, et al. (2004). "Direct proteomic mapping of the lung microvascular endothelial cell surface in vivo and in cell culture." *Nat Biotechnol* **22**(8): 985-92.

Edman, P. (1960). "Phenylthiohydantoins in protein analysis." *Ann N Y Acad Sci* **88**: 602-10.

Edman, P. and G. Begg (1967). "A protein sequenator." *Eur J Biochem* **1**(1): 80-91.

Emmert-Buck, M. R., R. F. Bonner, et al. (1996). "Laser capture microdissection." *Science* **274**(5289): 998-1001.

Eng, J. K., A. L. McCormack, et al. (1994). "An approach to correlate tandem mass-spectral data of peptides with amino-acid-sequences in a protein database." *J Am Soc Mass Spectrom* **11**: 976-989.

Fan, C., D. Oh, et al. (2006). "Concordance among Gene-Expression-Based Predictors for Breast Cancer." *The New England journal of medicine* **355**(6): 560.

Fang, R., D. A. Elias, et al. (2006). "Differential label-free quantitative proteomic analysis of Shewanella oneidensis cultured under aerobic and suboxic conditions by accurate mass and time tag approach." *Mol Cell Proteomics* **5**(4): 714-25.

Fenn, J. B., M. Mann, et al. (1989). "Electrospray ionization for mass spectrometry of large biomolecules." *Science* **246**(4926): 64-71.

Fidler, I. (1986). "Rationale and methods for the use of nude mice to study the biology and therapy of human cancer metastasis." *Cancer and Metastasis Reviews* **5**(1): 29-49.

Fidler, I. (1991). "Orthotopic implantation of human colon carcinomas into nude mice provides a valuable model for the biology and therapy of metastasis." *Cancer and Metastasis Reviews* **10**(3): 229-243.

Fidler, I. and I. Hart (1982). "Biological diversity in metastatic neoplasms: origins and implications." *Science* **217**(4564): 998.

Fidler, I. and M. Kripke (1977). "Metastasis results from preexisting variant cells within a malignant tumor." *Science* **197**(4306): 893.

Fidler, I., S. Naito, et al. (1990). "Orthotopic implantation is essential for the selection, growth and metastasis of human real cell cancer in nude mice." *Cancer and Metastasis Reviews* **9**(2): 149-165.

Fisher, G., S. Wellen, et al. (2001). "Induction and apoptotic regression of lung adenocarcinomas by regulation of a K-Ras transgene in the presence and absence of tumor suppressor genes." *Genes & Development* **15**(24): 3249-3262.

Florens, L., M. P. Washburn, et al. (2002). "A proteomic view of the Plasmodium falciparum life cycle." *Nature* **419**(6906): 520-6.

Fowlis, D. and A. Balmain (1993). "Oncogenes and tumour suppressor genes in transgenic mouse models of neoplasia." *Eur J Cancer* **29**(4): 638-45.

Furey, T. S., N. Cristianini, et al. (2000). "Support vector machine classification and validation of cancer tissue samples using microarray expression data." *Bioinformatics* **16**(10): 906-14.

Furukawa, T. (1993). "A novel" patient-like" treatment model of human pancreatic cancer constructed using orthotopic transplantation of histologically intact human tumor tissue in nude mice." *Cancer Research* **53**(13): 3070-3072.

Galski, H., M. Sullivan, et al. (1989). "Expression of a human multidrug resistance cDNA (MDR1) in the bone marrow of transgenic mice: resistance to daunomycin-induced leukopenia." *Molecular and Cellular Biology* **9**(10): 4357-4363.

Gao, X. Q., J. X. Han, et al. (2005). "Effect of NS398 on metastasis-associated gene expression in a human colon cancer cell line." *World J Gastroenterol* **11**(28): 4337-43.

Garber, M., O. Troyanskaya, et al. (2001). "Diversity of gene expression in adenocarcinoma of the lung." *Proceedings of the National Academy of Sciences* **98**(24): 13784-13789.

Gavin, A. C., M. Bosche, et al. (2002). "Functional organization of the yeast proteome by systematic analysis of protein complexes." *Nature* **415**(6868): 141-7.

Gazdar, A., V. Kurvari, et al. (1998). "Characterization of paired tumor and non-tumor cell lines established from patients with breast cancer." *Int J Cancer* **78**(6): 766-74.

Geer, L. Y., S. P. Markey, et al. (2004). "Open mass spectrometry search algorithm." *J Proteome Res* **3**(5): 958-64.

Gevaert, K. and J. Vandekerckhove (2000). "Protein identification methods in proteomics." *Electrophoresis* **21**(6): 1145-54.

Gilna, P. (2002). *Bioinformatics Workshop*. US National Institutes of Health.

Glaves, D. (1986). "Detection of circulating metastatic cells." *Prog Clin Biol Res* **212**: 151-67.

Golub, T., D. Slonim, et al. (1999). "Molecular Classification of Cancer: Class Discovery and Class Prediction by Gene Expression Monitoring." *Science* **286**(5439): 531-537.

Gordon, J. and F. Ruddle (1983). "Gene transfer into mouse embryos: production of transgenic mice by pronuclear injection." *Methods Enzymol* **101**: 411-33.

Gorg, A., C. Obermaier, et al. (2000). "The current state of two-dimensional electrophoresis with immobilized pH gradients." *Electrophoresis* **21**(6): 1037-53.

Greenes, R. A. and E. H. Shortliffe (1990). "Medical informatics. An emerging academic discipline and institutional priority." *Jama* **263**(8): 1114-20.

Gygi, S. P., B. Rist, et al. (1999). "Quantitative analysis of complex protein mixtures using isotope-coded affinity tags." *Nat Biotechnol* **17**(10): 994-9.

Haas, W., B. K. Faherty, et al. (2006). "Optimization and use of peptide mass measurement accuracy in shotgun proteomics." *Mol Cell Proteomics* **5**(7): 1326-37.

Hahn, M. W. and A. D. Kern (2005). "Comparative genomics of centrality and essentiality in three eukaryotic protein-interaction networks." *Mol Biol Evol* **22**(4): 803-6.

Han, J. D., N. Bertin, et al. (2004). "Evidence for dynamically organized modularity in the yeast protein-protein interaction network." *Nature* **430**(6995): 88-93.

Han, K. and Y. Byun (2004). "Three-dimensional visualization of protein interaction networks." *Comput Biol Med* **34**(2): 127-39.

Han, K., B. H. Ju, et al. (2004). "WebInterViewer: visualizing and analyzing molecular interaction networks." *Nucleic Acids Res* **32**(Web Server issue): W89-95.

Han, K., B. Park, et al. (2004). "HPID: the Human Protein Interaction Database." *Bioinformatics* **20**(15): 2466-70.

Hanahan, D. and R. Weinberg (2000). "The Hallmarks of Cancer." *Cell* **100**(1): 57-70.

Haux, R. (1997). "Aims and tasks of medical informatics." *Int J Med Inform* **44**(1): 9-20; discussion 39-44, 45-52, 61-6.

Hedley, D., M. Pintilie, et al. (2003). "Carbonic anhydrase IX expression, hypoxia, and prognosis in patients with uterine cervical carcinomas." *Clin Cancer Res* **9**(15): 5666-74.

Heppner, G. (1984). "Tumor heterogeneity." *Cancer Res* **44**(6): 2259-65.

Hermjakob, H., L. Montecchi-Palazzi, et al. (2004). "The HUPO PSI's molecular interaction format--a community standard for the representation of protein interaction data." *Nat Biotechnol* **22**(2): 177-83.

Hermjakob, H., L. Montecchi-Palazzi, et al. (2004). "IntAct: an open source molecular interaction database." *Nucleic Acids Res* **32 Database issue**: D452-5.

Hinton, G. E. (2000). "Computation by neural networks." *Nat Neurosci* **3 Suppl**: 1170.

Hinton, G. E. and R. R. Salakhutdinov (2006). "Reducing the dimensionality of data with neural networks." *Science* **313**(5786): 504-7.

Hirschman, L., A. Yeh, et al. (2005). "Overview of BioCreAtIvE: critical assessment of information extraction for biology." *BMC Bioinformatics* **6 Suppl 1**: S1.

Ho, Y., A. Gruhler, et al. (2002). "Systematic identification of protein complexes in Saccharomyces cerevisiae by mass spectrometry." *Nature* **415**(6868): 180-3.

Hoffmann, R. and A. Valencia (2005). "Implementing the iHOP concept for navigation of biomedical literature." *Bioinformatics* **21 Suppl 2**: ii252-ii258.

Howard, R. (1991). "Irradiated nude rat model for orthotopic human lung cancers." *Cancer Research* **51**(12): 3274-3280.

Howard, R., J. Mullen, et al. (1999). "Characterization of a highly metastatic, orthotopic lung cancer model in the nude rat." *Clinical and Experimental Metastasis* **17**(2): 157-162.

Huang, T. W., A. C. Tien, et al. (2004). "POINT: a database for the prediction of protein-protein interactions based on the orthologous interactome." *Bioinformatics* **(in press)**.

Hughes, T. R., M. J. Marton, et al. (2000). "Functional discovery via a compendium of expression profiles." *Cell* **102**(1): 109-26.

Iragne, F., M. Nikolski, et al. (2005). "ProViz: protein interaction visualization and exploration." *Bioinformatics* **21**(2): 272-4.

Irizarry, R. A., B. Hobbs, et al. (2003). "Exploration, normalization, and summaries of high density oligonucleotide array probe level data." *Biostatistics* **4**(2): 249-64.

Ivanov, Y. D., V. M. Govorun, et al. (2006). "Nanotechnologies in proteomics." *Proteomics* **6**(5): 1399-414.

Jackson, E., N. Willis, et al. (2001). Analysis of lung tumor initiation and progression using conditional expression of oncogenic K-ras, Cold Spring Harbor Laboratory Press.

Jaenisch, R. (1980). "Retroviruses and embryogenesis: microinjection of Moloney leukemia virus into midgestation mouse embryos." *Cell* **19**(18): ll.

Jaenisch, R., D. Jahner, et al. (1981). "Chromosomal position and activation of retroviral genomes inserted into the germ line of mice." *Cell* **24**(2): 519-529.

Jähner, D. and R. Jaenisch (1980). "Integration of Moloney leukaemia virus into the germ line of mice: correlation between site of integration and virus activation." *Nature* **287**: 456-458.

Jalbert, G. and A. Castonguay (1992). "Effects of NSAIDs on NNK-induced pulmonary and gastric tumorigenesis in A/J mice." *Cancer Lett* **66**(1): 21-8.

Jänne, P., J. Engelman, et al. (2005). "Epidermal Growth Factor Receptor Mutations in Non–Small-Cell Lung Cancer: Implications for Treatment and Tumor Biology." *Journal of Clinical Oncology* **23**(14): 3227-3234.

Jansen, R., D. Greenbaum, et al. (2002). "Relating whole-genome expression data with protein-protein interactions." *Genome Res* **12**(1): 37-46.

Jeong, H., S. P. Mason, et al. (2001). "Lethality and centrality in protein networks." *Nature* **411**(6833): 41-2.

Johnson, B. and P. Janne (1257). "Epidermal Growth Factor Receptor Mutations in Patients with Non-Small Cell Lung Cancer." *Cancer Research* **65**(17): 7525-7529.

Johnson, L., K. Mercer, et al. (2001). "Somatic activation of the K-ras oncogene causes early onset lung cancer in mice." *Nature* **410**(6832): 1111-6.

Johnston, M., J. Mullen, et al. (2001). Validation of an orthotopic model of human lung cancer with regional and systemic metastases, The Society of Thoracic Surgeons.

Jonsson, P. F. and P. A. Bates (2006). "Global topological features of cancer proteins in the human interactome." *Bioinformatics* **22**(18): 2291-7.

Jonsson, P. F., T. Cavanna, et al. (2006). "Cluster analysis of networks generated through homology: automatic identification of important protein communities involved in cancer metastasis." *BMC Bioinformatics* **7**(1): 2.

Jurisica, I. and J. Glasgow (2004). "Application of case-based reasoning in molecular biology." *Artificial Intelligence Magazine, Special issue on Bioinformatics* **25**(1): 85-95.

Jurisica, I., J. Mylopoulos, et al. (1998). "Case-based reasoning in IVF: prediction and knowledge mining." *Artif Intell Med* **12**(1): 1-24.

Kafatos, F. and T. Eisner (2004). Unification in the Century of Biology.

Kagolovsky, Y. and J. R. Moehr (2003). "Current status of the evaluation of information retrieval." *J Med Syst* **27**(5): 409-24.

Kagolovsky, Y. and J. R. Moehr (2003). "Terminological problems in information retrieval." *J Med Syst* **27**(5): 399-408.

Kahn, C. E., Jr. and P. N. Huynh (1996). "Knowledge representation for platform-independent structured reporting." *Proc AMIA Annu Fall Symp*: 478-82.

Kanehisa, M., S. Goto, et al. (2002). "The KEGG databases at GenomeNet." *Nucleic Acids Res* **30**(1): 42-6.

Karas, M. and F. Hillenkamp (1988). "Laser desorption ionization of proteins with molecular masses exceeding 10,000 daltons." *Anal Chem* **60**(20): 2299-301.

Kato, T., Y. Murata, et al. (2006). "Network-based de-noising improves prediction from microarray data." *BMC Bioinformatics* **7 Suppl 1**: S4.

Keller, A., A. I. Nesvizhskii, et al. (2002). "Empirical statistical model to estimate the accuracy of peptide identifications made by MS/MS and database search." *Anal Chem* **74**(20): 5383-92.

Kellett, C. F., A. L. Hart, et al. (1996). "Poor recall performance of journal-browsing doctors." *Lancet* **348**(9025): 479.

Kerbel, R., I. Cornil, et al. (1991). "Importance of orthotopic transplantation procedures in assessing the effects of transfected genes on human tumor growth and metastasis." *Cancer and Metastasis Reviews* **10**(3): 201-215.

Kerr, K. (2001). Pulmonary preinvasive neoplasia, Journal of Clinical Pathology.

Khleif SN, C. G. (1997). Animal models in drug development. *Cancer Medicine*. B. R. J. M. D. F. E. K. D. W. R. Holland JF. Baltimore, MD, Williams & Wilkins.

Kim, S. and C. Lee (1996). "Induction of benign and malignant pulmonary tumours in mice with benzo (a) pyrene." *Anticancer Res* **16**(1): 465-70.

King, A. D., N. Przulj, et al. (2004). "Protein complex prediction via cost-based clustering." *Bioinformatics* **20**(17): 3013-20.

Kislinger, T., B. Cox, et al. (2006). "Global survey of organ and organelle protein expression in mouse: combined proteomic and transcriptomic profiling." *Cell* **125**(1): 173-86.

Kislinger, T. and I. Jurisica (2006). "Proteomics and bioinformatics in biomedical research." *Cancer Genomics and Proteomics* **3**(1): 11-28.

Kislinger, T., K. Rahman, et al. (2003). "PRISM, a generic large scale proteomic investigation strategy for mammals." *Mol Cell Proteomics* **2**(2): 96-106.

Kobourov, S. G. and K. Wampler (2005). "Non-Euclidean spring embedders." *IEEE Trans Vis Comput Graph* **11**(6): 757-67.

Kohonen, T. (1995). *Self-organizing maps*. Berlin ; New York, Springer.

Koller, A., M. P. Washburn, et al. (2002). "Proteomic survey of metabolic pathways in rice." *Proc Natl Acad Sci U S A* **99**(18): 11969-74.

Kotlyar, M. and I. Jurisica (2006). "Predicting protein-protein interactions by association mining." *Information Systems Frontiers* **8**: 37-47.

Kozaki, K., K. Koshikawa, et al. "Multi-faceted analyses of a highly metastatic human lung cancer cell line NCI-H460-LNM35 suggest mimicry of inflammatory cells in metastasis."

Kozaki, K., O. Miyaishi, et al. (2000). Establishment and Characterization of a Human Lung Cancer Cell Line NCI-H460-LNM35 with Consistent Lymphogenous Metastasis via Both Subcutaneous and Orthotopic Propagation 1.

Kramer, R. and D. Cohen (2004). "Functional genomics to new drug targets." *Nature Reviews Drug Discovery* **3**(11): 965-972.

Kraus-Berthier, L., M. Jan, et al. (2000). Histology and Sensitivity to Anticancer Drugs of Two Human Non-Small Cell Lung Carcinomas Implanted in the Pleural Cavity of Nude Mice.

Kuo, T., T. Kubota, et al. (1992). "Orthotopic reconstitution of human small-cell lung carcinoma after intravenous transplantation in SCID mice." *Anticancer Res* **12**(5): 1407-10.

Kuo, T., T. Kubota, et al. (1993). "Site-specific chemosensitivity of human small-cell lung carcinoma growing orthotopically compared to subcutaneously in SCID mice: the importance of orthotopic models to obtain relevant drug evaluation data." *Anticancer Res* **13**(3): 627-30.

Lau, S. K., P. C. Boutros, et al. (2007). "Minimal gene expression classifiers for molecular staging of early-stage non-small cell lung cancer patients." *Submitted*.

Lehmann, T. M., T. Aach, et al. (2006). "Sensor, signal, and image informatics - state of the art and current topics." *Methods Inf Med* **45 Suppl 1**: 57-67.

Lehmann, T. M., M. O. Guld, et al. (2004). "Content-based image retrieval in medical applications." *Methods Inf Med* **43**(4): 354-61.

Li, W., L. Wen, et al. (2006). "RepairNET: A bioinformatics toolbox for functional exploration of DNA damage response." *J Cell Physiol*.

Li, X. J., E. C. Yi, et al. (2005). "A software suite for the generation and comparison of peptide arrays from sets of data collected by liquid chromatography-mass spectrometry." *Mol Cell Proteomics* **4**(9): 1328-40.

Lin, W., T. Pretlow, et al. (1990). "Bacterial lacZ gene as a highly sensitive marker to detect micrometastasis formation during tumor progression." *Cancer Res* **50**(9): 2808-17.

Link, A. J., J. Eng, et al. (1999). "Direct analysis of protein complexes using mass spectrometry." *Nat Biotechnol* **17**(7): 676-82.

Liotta, L., J. Kleinerman, et al. (1974). "Quantitative relationships of intravascular tumor cells, tumor vessels, and pulmonary metastases following tumor implantation." *Cancer Res* **34**(5): 997-1004.

Listgarten, J., S. Damaraju, et al. (2004). "Predictive models for breast cancer susceptibility from multiple single nucleotide polymorphisms." *Clin Cancer Res* **10**(8): 2725-37.

Liu, H., R. G. Sadygov, et al. (2004). "A model for random sampling and estimation of relative protein abundance in shotgun proteomics." *Anal Chem* **76**(14): 4193-201.

Liu, Y., S. B. Navathe, et al. (2005). "Text mining biomedical literature for discovering gene-to-gene relationships: a comparative study of algorithms." *IEEE/ACM Trans Comput Biol Bioinform* **2**(1): 62-76.

Livingood, L. (1986). "Tumors in the mouse." *Johns Hopkins Bulletin* **66**(67): 177.

Lu, H., B. Shi, et al. (2006). "Integrated analysis of multiple data sources reveals modular structure of biological networks." *Biochem Biophys Res Commun* **345**(1): 302-9.

Lynch, T., D. Bell, et al. (2004). Activating Mutations in the Epidermal Growth Factor Receptor Underlying Responsiveness of Non-Small-Cell Lung Cancer to Gefitinib.

MacCoss, M. J., C. C. Wu, et al. (2002). "Probability-based validation of protein identifications using a modified SEQUEST algorithm." *Anal Chem* **74**(21): 5593-9.

Macura, R. T., K. J. Macura, et al. (1994). "Computerized case-based instructional system for computed tomography and magnetic resonance imaging of brain tumors." *Invest Radiol* **29**(4): 497-506.

Majoros, W. H., G. M. Subramanian, et al. (2003). "Identification of key concepts in biomedical literature using a modified Markov heuristic." *Bioinformatics* **19**(3): 402-7.

Malkinson, A. (1989). "The genetic basis of susceptibility to lung tumors in mice." *Toxicology* **54**(3): 241-71.

Malkinson, A. (1992). "Primary lung tumors in mice: an experimentally manipulable model of human adenocarcinoma." *Cancer Research* **52**(9): 2670-2676.

Manzotti, C., R. Audisio, et al. (1993). "Importance of orthotopic implantation for human tumors as model systems: relevance to metastasis and invasion." *Clinical and Experimental Metastasis* **11**(1): 5-14.

Mao, W. and W. W. Chu (2002). "Free-text medical document retrieval via phrase-based vector space model." *Proc AMIA Symp*: 489-93.

Marandola, P., A. Bonghi, et al. (2004). "Molecular biology and the staging of prostate cancer." *Ann N Y Acad Sci* **1028**: 294-312.

Maronpot, R., R. Palmiter, et al. (1991). "Pulmonary carcinogenesis in transgenic mice." *Exp Lung Res* **17**(2): 305-20.

Maslov, S. and K. Sneppen (2002). "Specificity and stability in topology of protein networks." *Science* **296**(5569): 910-3.

Mattern, J., M. Bak, et al. (1988). "Human tumor xenografts as model for drug testing." *Cancer and Metastasis Reviews* **7**(3): 263-284.

McCann, J. (2000). "Molecular markers may improve colon cancer staging, screening." *J Natl Cancer Inst* **92**(13): 1039-40.

McLemore, T. (1987). "Novel intrapulmonary model for orthotopic propagation of human lung cancers in athymic nude mice." *Cancer Research* **47**(19): 5132-5140.

McLemore, T. (1988). "Comparison of intrapulmonary, percutaneous intrathoracic, and subcutaneous models for the propagation of human pulmonary and nonpulmonary cancer cell lines in athymic nude mice." *Cancer Research* **48**(10): 2880-2886.

Meistermann, H., J. L. Norris, et al. (2006). "Biomarker discovery by imaging mass spectrometry: Transthyretin is a biomarker for gentamicin-induced nephrotoxicity in rat." *Mol Cell Proteomics*.

Meuwissen, R., S. Linn, et al. (2001). "Mouse model for lung tumorigenesis through Cre/lox controlled sporadic activation of the K-Ras oncogene."

Mewes, H. W., D. Frishman, et al. (2002). "MIPS: a database for genomes and protein sequences." *Nucleic Acids Res* **30**(1): 31-4.

Milo, R., S. Shen-Orr, et al. (2002). "Network motifs: simple building blocks of complex networks." *Science* **298**(5594): 824-7.

Miotto, O., T. W. Tan, et al. (2005). "Supporting the curation of biological databases with reusable text mining." *Genome Inform* **16**(2): 32-44.

Miyake, M., M. Adachi, et al. (1999). "A novel molecular staging protocol for non-small cell lung cancer." *Oncogene* **18**(14): 2397-404.

Miyoshi, T., K. Kondo, et al. (2000). "SCID mouse lymphogenous metastatic model of human lung cancer constructed using orthotopic inoculation of cancer cells." *Anticancer Res* **20**(1A): 161-3.

Moody, T., J. Leyton, et al. (2001). "Indomethacin reduces lung adenoma number in A/J mice." *Anticancer Res* **21**(3B): 1749-55.

Moon, R., K. Rao, et al. (1992). "Hamster lung cancer model of carcinogenesis and chemoprevention." *Adv Exp Med Biol* **320**: 55-61.

Morikawa, K. (1988). "In vivo selection of highly metastatic cells from surgical specimens of different primary human colon carcinomas implanted into nude mice." *Cancer Research* **48**(7): 1943-1948.

Morse, M. (1991). "Structure-activity relationships for inhibition of 4-(methylnitrosamino)-1-(3-pyridyl)-1-butanone lung tumorigenesis by arylalkyl isothiocyanates in A/J mice." *Cancer Research* **51**(7): 1846-1850.

Motakis, E. S., G. P. Nason, et al. (2006). "Variance stabilization and normalization for one-color microarray data using a data-driven multiscale approach." *Bioinformatics* **22**(20): 2547-53.

Motamed-Khorasani, A., I. Jurisica, et al. (2007). "Differentially androgen-modulated genes in ovarian epithelial cells from BRCA mutation carriers and control patients predict ovarian cancer survival and disease progression." *Oncogene* **26**(2): 198-214.

Mountain, C. (1987). "The new International Staging System for Lung Cancer." *Surg Clin North Am* **67**(5): 925-35.

Mountain, C. (1997). "Regional lymph node classification for lung cancer staging." *Chest* **111**(6): 1718-1723.

Mountain, C., D. Carr, et al. (1974). "A System For The Clinical Staging of Lung Cancer." *American Journal of Roentgenology* **120**(1): 130-138.

Mulvin, D., R. Howard, et al. (1993). "Secondary screening system for preclinical testing of human lung cancer therapies." *JNCI Cancer Spectrum* **84**: 31-37.

Naito, S. (1986). "Growth and metastasis of tumor cells isolated from a human renal cell carcinoma implanted into different organs of nude mice." *Cancer Research* **46**(8): 4109-4115.

Nesvizhskii, A. I., A. Keller, et al. (2003). "A statistical model for identifying proteins by tandem mass spectrometry." *Anal Chem* **75**(17): 4646-58.

Neuvial, P., P. Hupe, et al. (2006). "Spatial normalization of array-CGH data." *BMC Bioinformatics* **7**: 264.

Nicolson, G. (1984). "Generation of phenotypic diversity and progression in metastatic tumor cells." *Cancer and Metastasis Reviews* **3**(1): 25-42.

Nicolson, G. (1987). "Tumor cell instability, diversification, and progression to the metastatic phenotype: from oncogene to oncofetal expression." *Cancer Research* **47**(6): 1473-1487.

Niijima, S. and S. Kuhara (2005). "Multiclass molecular cancer classification by kernel subspace methods with effective kernel parameter selection." *J Bioinform Comput Biol* **3**(5): 1071-88.

Nikkila, J., P. Toronen, et al. (2002). "Analysis and visualization of gene expression data using self-organizing maps." *Neural Netw* **15**(8-9): 953-66.

Ogata, H., S. Goto, et al. (1999). "KEGG: Kyoto Encyclopedia of Genes and Genomes." *Nucleic Acids Res* **27**(1): 29-34.

Omenn, G. S., D. J. States, et al. (2005). "Overview of the HUPO Plasma Proteome Project: results from the pilot phase with 35 collaborating laboratories and multiple analytical groups, generating a core dataset of 3020 proteins and a publicly-available database." *Proteomics* **5**(13): 3226-45.

Ong, L. S., B. Shepherd, et al. (1997). "The Colorectal Cancer Recurrence Support (CARES) System." *Artif Intell Med* **11**(3): 175-88.

Ong, S. E., B. Blagoev, et al. (2002). "Stable isotope labeling by amino acids in cell culture, SILAC, as a simple and accurate approach to expression proteomics." *Mol Cell Proteomics* **1**(5): 376-86.

Ong, S. E. and M. Mann (2005). "Mass spectrometry-based proteomics turns quantitative." *Nat Chem Biol* **1**(5): 252-62.

Ono, M., M. Shitashige, et al. (2006). "Label-free quantitative proteomics using large peptide data sets generated by nanoflow liquid chromatography and mass spectrometry." *Mol Cell Proteomics* **5**(7): 1338-47.

Otasek, D., K. R. Brown, et al. (2006). *Confirming protein-protein interactions by text mining*. Society for Industrual and Applied Mathematics (SIAM) Conference on Data Mining; Text mining, Bethesda, Maryland, SIAM.

Oyama, T., K. Kitano, et al. (2002). "Extraction of knowledge on protein-protein interaction by association rule discovery." *Bioinformatics* **18**(5): 705-14.

Paez, J., P. Janne, et al. (2004). EGFR Mutations in Lung Cancer: Correlation with Clinical Response to Gefitinib Therapy, American Association for the Advancement of Science.

Paget, S. (1889). "The distribution of secondary growths in cancer of the breast." *Lancet* **1**(6): 571.

Paik, S., S. Shak, et al. (2004). A Multigene Assay to Predict Recurrence of Tamoxifen-Treated, Node-Negative Breast Cancer.

Palakal, M., M. Stephens, et al. (2002). "A multi-level text mining method to extract biological relationships." *Proc IEEE Comput Soc Bioinform Conf* **1**: 97-108.

Pant, D. K. and A. Ghosh (2006). "A systems biology approach for the study of cumulative oncogenes with applications to the MAPK signal transduction pathway." *Biophys Chem* **119**(1): 49-60.

Pantazi, S. V., J. F. Arocha, et al. (2004). "Case-based medical informatics." *BMC Med Inform Decis Mak* **4**(1): 19.

Pantazi, S. V., A. Kushniruk, et al. (2006). "The usability axiom of medical information systems." *Int J Med Inform* **75**(12): 829-39.

Pare, G. and M. C. Trudel (2007). "Knowledge barriers to PACS adoption and implementation in hospitals." *Int J Med Inform* **76**(1): 22-33.

Peng, J., J. E. Elias, et al. (2003). "Evaluation of multidimensional chromatography coupled with tandem mass spectrometry (LC/LC-MS/MS) for large-scale protein analysis: the yeast proteome." *J Proteome Res* **2**(1): 43-50.

Peri, S., J. D. Navarro, et al. (2004). "Human protein reference database as a discovery resource for proteomics." *Nucleic Acids Res* **32**(Database issue): D497-501.

Perkins, D. N., D. J. Pappin, et al. (1999). "Probability-based protein identification by searching sequence databases using mass spectrometry data." *Electrophoresis* **20**(18): 3551-67.

Petricoin, E. F., A. M. Ardekani, et al. (2002). "Use of proteomic patterns in serum to identify ovarian cancer." *Lancet* **359**(9306): 572-7.

Pirooznia, M. and Y. Deng (2006). "SVM Classifier - a comprehensive java interface for support vector machine classification of microarray data." *BMC Bioinformatics* **7 Suppl 4**: S25.

Potti, A., S. Mukherjee, et al. (2006). "A Genomic Strategy to Refine Prognosis in Early-Stage Non-Small-Cell Lung Cancer." *New England Journal of Medicine* **355**(6): 570.

Povlsen, C. and J. Rygaard (1971). "Heterotransplantation of human adenocarcinomas of the colon and rectum to the mouse mutant Nude. A study of nine consecutive transplantations." *Acta Pathol Microbiol Scand [A* **79**(2): 159-69.

Price, J. (1994). "Analyzing the metastatic phenotype." *J Cell Biochem* **56**(1): 16-22.

Przulj, N., D. G. Corneil, et al. (2004). "Modeling interactome: scale-free or geometric?" *Bioinformatics* **20**(18): 3508-15.

Przulj, N., D. G. Corneil, et al. (2006). "Efficient estimation of graphlet frequency distributions in protein-protein interaction networks." *Bioinformatics* **22**(8): 974-80.

Przulj, N., D. A. Wigle, et al. (2004). "Functional topology in a network of protein interactions." *Bioinformatics* **20**(3): 340-8.

Quackenbush, J. (2002). "Microarray data normalization and transformation." *Nat Genet* **32 Suppl**: 496-501.

Rabbee, N. and T. P. Speed (2006). "A genotype calling algorithm for affymetrix SNP arrays." *Bioinformatics* **22**(1): 7-12.

Radulovic, D., S. Jelveh, et al. (2004). "Informatics platform for global proteomic profiling and biomarker discovery using liquid chromatography-tandem mass spectrometry." *Mol Cell Proteomics* **3**(10): 984-97.

Rapp, U. and G. Todaro (1980). "Generation of Oncogenic Mouse type C Viruses: in vitro Selection of Carcinoma-Inducing Variants." *Proceedings of the National Academy of Sciences* **77**(1): 624-628.

Rashidi, B., M. Yang, et al. (2000). "A highly metastatic Lewis lung carcinoma orthotopic green fluorescent protein model." *Clinical and Experimental Metastasis* **18**(1): 57-60.

Ravasz, E., A. L. Somera, et al. (2002). "Hierarchical organization of modularity in metabolic networks." *Science* **297**(5586): 1551-5.

Reyzer, M. L. and R. M. Caprioli (2005). "MALDI mass spectrometry for direct tissue analysis: a new tool for biomarker discovery." *J Proteome Res* **4**(4): 1138-42.

Roboz, J. (2005). "Mass spectrometry in diagnostic oncoproteomics." *Cancer Invest* **23**(5): 465-78.

Rossille, D., J. F. Laurent, et al. (2005). "Modelling a decision-support system for oncology using rule-based and case-based reasoning methodologies." *Int J Med Inform* **74**(2-4): 299-306.

Russell, P., I. Shon, et al. (1991). "Growth and metastasis of human bladder cancer xenografts in the bladder of nude rats." *Urological Research* **19**(4): 207-213.

Sandmoller, A. (1995). "A transgenic mouse model for lung adenocarcinoma." *Cell Growth and Differentiation* **6**(1): 97-103.

Schirmer, E. C., L. Florens, et al. (2003). "Nuclear membrane proteins with potential disease links found by subtractive proteomics." *Science* **301**(5638): 1380-2.

Schmidt, A., J. Kellermann, et al. (2005). "A novel strategy for quantitative proteomics using isotope-coded protein labels." *Proteomics* **5**(1): 4-15.

Schuller, H. (1985). "Inhibition of N-nitrosodiethylamine-induced respiratory tract carcinogenesis by piperonylbutoxide in hamsters." *Cancer Research* **45**(6): 2807-2812.

Schuster, J. (1993). "Intraarterial therapy of human glioma xenografts in athymic rats using 4-hydroperoxycyclophosphamide." *Cancer Research* **53**(10): 2338-2343.

Scott, M. S., S. J. Calafell, et al. (2005). "Refining protein subcellular localization." *PLoS Comput Biol* **1**(6): e66.

Scott, M. S., T. Perkins, et al. (2005). "Identifying regulatory subnetworks for a set of genes." *Mol Cell Proteomics* **4**(5): 683-92.

Seiden-Long, I. M., K. R. Brown, et al. (2006). "Transcriptional targets of hepatocyte growth factor signaling and Ki-ras oncogene activation in colorectal cancer." *Oncogene* **25**(1): 91-102.

Sen, T. Z., A. Kloczkowski, et al. (2006). "Functional clustering of yeast proteins from the protein-protein interaction network." *BMC Bioinformatics* **7**: 355.

Shannon, P., A. Markiel, et al. (2003). "Cytoscape: a software environment for integrated models of biomolecular interaction networks." *Genome Res* **13**(11): 2498-504.

Shepherd, F., P. JR, et al. (2005). "Erlotinib in previously treated non-small-cell lung cancer." *The New England journal of medicine* **353**(2): 123-132.

Shi, L., L. H. Reid, et al. (2006). "The MicroArray Quality Control (MAQC) project shows inter- and intraplatform reproducibility of gene expression measurements." *Nat Biotechnol* **24**(9): 1151-61.

Shimkin, M. and G. Stoner (1975). "Lung tumors in mice: application to carcinogenesis bioassay." *Adv Cancer Res* **21**: 1-58.

Shockett, P. and D. Schatz (1996). Diverse strategies for tetracycline-regulated inducible gene expression.

Shoemaker RH, M. T., Abbott BJ, et al. . (1988). Human tumor xenograft models for use with an in vitro-based, disease-oriented antitumor drug screening program. *Human Tumor xenografts in Anticancer Drug Development*. B. W. M. P. a. H. P. (eds). Berlin, Springer-Verlag.

Shortliffe, E. H. and E. J. Sondik (2006). "The public health informatics infrastructure: anticipating its role in cancer." *Cancer Causes Control* **17**(7): 861-9.

Simon, R. M., A. Lam, et al. (2007). "Analysis of Gene Expression Data Using BRB-Array Tools." *Cancer Informatics* **3**: 11-17.

Singletary, S., C. Allred, et al. (2002). "Revision of the American Joint Committee on Cancer Staging System for Breast Cancer." *Journal of Clinical Oncology* **20**(17): 3628-3636.

Smyth, G. K. and T. Speed (2003). "Normalization of cDNA microarray data." *Methods* **31**(4): 265-73.

Sneppen, K., P. Bak, et al. (1995). "Evolution as a self-organized critical phenomenon." *Proc Natl Acad Sci U S A* **92**(11): 5209-13.

Sobin LH, W. C. e. (2002). *TNM Classification of Malignant Tumours, 6th Edition*, Wiley.

Society, A. C. (2006). "Cancer Facts and Figures 2006." from http://www.cancer.org/downloads/STT/.

Soriano, P. and R. Jaenisch (1986). "Retroviruses as probes for mammalian development: allocation of cells to the somatic and germ cell lineages." *Cell* **46**(1): 19-29.

Spinosa, E. J. and A. C. Carvalho (2005). "Support vector machines for novel class detection in Bioinformatics." *Genet Mol Res* **4**(3): 608-15.

States, D. J., G. S. Omenn, et al. (2006). "Challenges in deriving high-confidence protein identifications from data gathered by a HUPO plasma proteome collaborative study." *Nat Biotechnol* **24**(3): 333-8.

Steel, G., V. Courtenay, et al. (1983). "The response to chemotherapy of a variety of human tumour xenografts." *Br J Cancer* **47**(1): 1-13.

Steele, V., R. Moon, et al. (1994). "Preclinical efficacy evaluation of potential chemopreventive agents in animal carcinogenesis models: methods and results from the NCI Chemoprevention Drug Development Program." *J Cell Biochem Suppl* **20**: 32-54.

Steen, H. and M. Mann (2004). "The ABC's (and XYZ's) of peptide sequencing." *Nat Rev Mol Cell Biol* **5**(9): 699-711.

Stefancic, H. and V. Zlatic (2005). ""Winner takes it all": strongest node rule for evolution of scale-free networks." *Phys Rev E Stat Nonlin Soft Matter Phys* **72**(3 Pt 2): 036105.

Stern, H. and L. Zon (2003). "Cancer genetics and drug discovery in the zebrafish." *Nat Rev Cancer* **3**(7): 533-539.

Stoner, G. (1991). "Lung tumors in strain A mice as a bioassay for carcinogenicity of environmental chemicals." *Exp Lung Res* **17**(2): 405-23.

Suda, Y., S. Aizawa, et al. (1987). "Driven by the same Ig enhancer and SV40 T promoter ras induced lung adenomatous tumors, myc induced pre-B cell lymphomas and SV40 large T gene a variety of tumors in transgenic mice." *EMBO J* **6**(13): 4055-4065.

Sultan, M., D. A. Wigle, et al. (2002). "Binary tree-structured vector quantization approach to clustering and visualizing microarray data." *Bioinformatics* **18 Suppl 1**: S111-S119.

Takeuchi, F. (2005). "Effectiveness of vaccination strategies for infectious diseases according to human contact networks." *Jpn J Infect Dis* **58**(6): S16-7.

Tamayo, P., D. Slonim, et al. (1999). "Interpreting patterns of gene expression with self-organizing maps: methods and application to hematopoietic differentiation." *Proc Natl Acad Sci U S A* **96**(6): 2907-12.
Tanaka, K., H. Waki, et al. (1988). "Protein and polymer analysis up to m/z 100000 by laser ionization time-of-flight mass spectrometry." *Rapid Communications in Mass Spectrometry* **2**(8): 151-153.
Thatcher, N., A. Chang, et al. (2005). "Gefitinib plus best supportive care in previously treated patients with refractory advanced non-small-cell lung cancer: results from a randomised, placebo-controlled, multicentre study (Iressa Survival Evaluation in Lung Cancer)." *The Lancet* **366**(9496): 1527-1537.
Thomas, H. and F. Balkwill (1995). "Assessing new anti-tumour agents and strategies in oncogene transgenic mice." *Cancer and Metastasis Reviews* **14**(2): 91-95.
Topinka, C. and C. R. Shyu (2006). "Predicting cancer interaction networks using text-mining and structure understanding." *AMIA Annu Symp Proc*: 1123.
Toronen, P., M. Kolehmainen, et al. (1999). "Analysis of gene expression data using self-organizing maps." *FEBS Lett* **451**(2): 142-6.
Tsao, M., A. Sakurada, et al. (2005). Erlotinib in Lung Cancer-Molecular and Clinical Predictors of Outcome.
Ullrich, A. and J. Schlessinger (1990). "Signal transduction by receptors with tyrosine kinase activity." *Cell* **61**(2): 203-212.
van Weerden, W. and J. Romijn (2000). "Use of nude mouse xenograft models in prostate cancer research." *The Prostate* **43**(4): 263-271.
Venter, J., M. Adams, et al. (2001). "The Sequence of the Human Genome." *Science* **291**(5507): 1304-1351.
Verkoeijen, P. P., R. M. Rikers, et al. (2004). "Case representation by medical experts, intermediates and novices for laboratory data presented with or without a clinical context." *Med Educ* **38**(6): 617-27.
Vogelstein, B. and K. W. Kinzler (2004). "Cancer genes and the pathways they control." *Nat Med* **10**(8): 789-99.
von Mering, C., M. Huynen, et al. (2003). "STRING: a database of predicted functional associations between proteins." *Nucleic Acids Res* **31**(1): 258-61.
Wachi, S., K. Yoneda, et al. (2005). "Interactome-transcriptome analysis reveals the high centrality of genes differentially expressed in lung cancer tissues." *Bioinformatics* **21**(23): 4205-8.
Wang, G., W. W. Wu, et al. (2006). "Label-free protein quantification using LC-coupled ion trap or FT mass spectrometry: Reproducibility, linearity, and application with complex proteomes." *J Proteome Res* **5**(5): 1214-23.
Wang, X., X. Fu, et al. (1992). "A new patient-like metastatic model of human lung cancer constructed orthotopically with intact tissue via thoracotomy in immunodeficient mice." *Int J Cancer* **51**(6): 992-5.
Wang, X., H. He, et al. (2006). "NMPP: a user-customized NimbleGen microarray data processing pipeline." *Bioinformatics* **22**(23): 2955-7.

Washburn, M. P., D. Wolters, et al. (2001). "Large-scale analysis of the yeast proteome by multidimensional protein identification technology." *Nat Biotechnol* **19**(3): 242-7.

Wasinger, V. C., S. J. Cordwell, et al. (1995). "Progress with gene-product mapping of the Mollicutes: Mycoplasma genitalium." *Electrophoresis* **16**(7): 1090-4.

Wattenberg, L. (1972). "Inhibition of carcinogenic and toxic effects of polycyclic hydrocarbons by phenolic antioxidants and ethoxyquin." *J Natl Cancer Inst* **48**(5): 1425-30.

Wigle, D., I. Jurisica, et al. (2002). Molecular Profiling of Non-Small Cell Lung Cancer and Correlation with Disease-free Survival 1.

Wigle, D. A., I. Jurisica, et al. (2002). "Molecular Profiling of Non-Small Cell Lung Cancer and Correlation with Disease-free Survival." *Cancer Res* **62**(11): 3005-3008.

Wikenheiser, K. (1992). "Simian virus 40 large T antigen directed by transcriptional elements of the human surfactant protein C gene produces pulmonary adenocarcinomas in transgenic mice." *Cancer Research* **52**(19): 5342-5352.

Wilmanns, C., D. Fan, et al. (1992). "Orthotopic and ectopic organ environments differentially influence the sensitivity of murine colon carcinoma cells to doxorubicin and 5-fluorouracil." *Int J Cancer* **52**(1): 98-104.

Wistuba, I., D. Bryant, et al. (1999). Comparison of Features of Human Lung Cancer Cell Lines and Their Corresponding Tumors 1.

Wolters, D. A., M. P. Washburn, et al. (2001). "An automated multidimensional protein identification technology for shotgun proteomics." *Anal Chem* **73**(23): 5683-90.

Wu, C. C., M. J. MacCoss, et al. (2003). "A method for the comprehensive proteomic analysis of membrane proteins." *Nat Biotechnol* **21**(5): 532-8.

Wu, C. C., M. J. MacCoss, et al. (2004). "Organellar proteomics reveals Golgi arginine dimethylation." *Mol Biol Cell* **15**(6): 2907-19.

Wuchty, S. and E. Almaas (2005). "Evolutionary cores of domain co-occurrence networks." *BMC Evol Biol* **5**(1): 24.

Wuchty, S., A. L. Barabasi, et al. (2006). "Stable evolutionary signal in a Yeast protein interaction network." *BMC Evol Biol* **6**(1): 8.

Xenarios, I., D. W. Rice, et al. (2000). "DIP: the database of interacting proteins." *Nucleic Acids Res* **28**(1): 289-91.

Xi, L., W. Gooding, et al. (2006). "Identification of mRNA markers for molecular staging of lymph nodes in colorectal cancer." *Clin Chem* **52**(3): 520-3.

Yan, J. X., A. T. Devenish, et al. (2002). "Fluorescence two-dimensional difference gel electrophoresis and mass spectrometry based proteomic analysis of Escherichia coli." *Proteomics* **2**(12): 1682-98.

Yang, Y. H., S. Dudoit, et al. (2002). "Normalization for cDNA microarray data: a robust composite method addressing single and multiple slide systematic variation." *Nucleic Acids Res* **30**(4): e15.

Yardy, G., S. McGregor, et al. (2005). "The molecular staging of prostate cancer." *BJU Int* **95**(6): 907.

Yates, J. R., D. Cociorva, et al. (2006). "Performance of a linear ion trap-Orbitrap hybrid for peptide analysis." *Anal Chem* **78**(2): 493-500.

Yeh, A. S., L. Hirschman, et al. (2003). "Evaluation of text data mining for database curation: lessons learned from the KDD Challenge Cup." *Bioinformatics* **19 Suppl 1**: i331-9.

Yu, H. and M. Gerstein (2006). "Genomic analysis of the hierarchical structure of regulatory networks." *Proc Natl Acad Sci U S A*.

Yu, H., K. Nguyen, et al. (2007). "Positional artifacts in microarrays: experimental verification and construction of COP, an automated detection tool." *Nucleic Acids Res* **35**(2): e8.

Yuan, D. S. and R. A. Irizarry (2006). "High-resolution spatial normalization for microarrays containing embedded technical replicates." *Bioinformatics* **22**(24): 3054-60.

Zanzoni, A., L. Montecchi-Palazzi, et al. (2002). "MINT: a Molecular INTeraction database." *FEBS Lett* **513**(1): 135-40.

Zhao, B. (2000). "TRANSGENIC MOUSE MODELS FOR LUNG CANCER." *Experimental Lung Research* **26**(8): 567-579.

Zheng, P. Z., K. K. Wang, et al. (2005). "Systems analysis of transcriptome and proteome in retinoic acid/arsenic trioxide-induced cell differentiation/apoptosis of promyelocytic leukemia." *Proc Natl Acad Sci U S A* **102**(21): 7653-8.

Zhu, C.-Q., I. Jurisica, et al. (2007). "Five-gene prognostic classifiers in NSCLC." *Submitted*.

Zubrod, C. (1972). "Chemical Control of Cancer." *Proceedings of the National Academy of Sciences* **69**(4): 1042-1047.

Index

2
2-DE, 64, 75, 153, 157

A
association mining, 135, 170

B
BTSVQ, 48, 136, 140

C
case-based reasoning, 162, 169, 175
cell lines, 16, 22, 23, 33, 34, 36, 37, 39, 41, 43, 166, 172
CID, 62, 152
Ciphergen Biosystems., 76, 155
Collision induced dissociation, 152

D
Detection depth, 153
DICOM, 95, 105, 106

E
Edman sequencing, 71, 153
e-health, 165
Electrospray ionization, 61, 153, 165
ESI, 60, 61, 153
Evidence-based medicine, 153
Expression proteomics, 59, 156

F
Functional proteomics, 60, 156

G
Gene expression, 50
Gene Ontology, 66, 93, 153, 156, 161, 162

H
HIPAA, 92, 102, 104, 107, 110, 111
HL7, 90, 91, 92, 93, 94, 95, 96, 97, 98, 99, 102, 104, 106, 107, 110

I
ICAT, 74, 154
ICPL, 74, 154
Information-based medicine, 109, 153
Isotope coded affinity tags, 154

K
knowledge discovery, 13, 131, 137, 147, 152
knowledge management, 2, 85, 129, 130, 131

L
Laser capture microdissection, 77, 154, 165
LC-MS, 61, 73, 74, 80, 82, 154, 156, 174

Liquid chromatography mass
spectrometry, 154

M

machine learning, 85, 125, 134, 135,
151
MALDI-TOF-MS, 67, 70, 76, 154
Mascot, 70, 72, 154
Mass spectrometer, 154
Matrix Science, 72, 154
Metabolomics, 59, 155
Microarray, 59, 93, 97, 106, 175
Molecular staging, 161
MS/MS, 62, 70, 71, 76, 82, 155, 169
MudPIT, 63, 64, 66, 72, 155, 156
Multidimensional protein
identification technology, 63, 64,
155
Multidimensional protein
identification technology -
MudPIT, 63

N

NCBI, 72, 116, 155

O

OLAP, 120, 121, 122, 123, 124
OMSSA, 72, 155
Orthotopic, 33, 34, 165, 166, 170, 179

P

Peptide mass fingerprint, 70, 155
PPI, 132, 137, 140
PRISM, 66, 156, 170
protein-protein interactions, 60, 85,
125, 132, 141, 156, 168, 170, 174
Proteome, 72, 81, 156, 158, 163,
166, 173, 174, 175, 178
Proteomics, 59, 60, 80, 93, 97, 98,
106, 107, 156, 162, 163, 164, 165,
167, 168, 170, 171, 172, 173, 174,
175, 176, 179
Proteomics investigation strategy for
Mammals, 156

Proteomics of posttranslational
modifications, 60, 156

S

SELDI-TOF-MS, 76, 77, 154
self-organizing maps, 48, 49, 125,
135, 163, 173, 178
SEQUEST, 71, 72, 157, 171
SILAC, 74, 157, 174
Silica-bead perfusion technique,
157
snap-frozen, 16, 17, 18, 21
SOMs, 48, 49, 125
Stable Isotope Labeling with Amino
acids in Cell culture, 157
Structural proteomics, 60, 156
Surface-enhanced laser desorption
ionization time-of-flight mass
spectrometry, 154
systems biology, 3, 26, 59, 130, 149,
161, 174

T

Tandem mass spectrometry, 155
Text mining, 137, 171, 174
Thermo Fisher Scientific, 62, 71
tissue microarray, 17
TNM, 4, 5, 45, 46, 49, 50, 52, 54,
164, 177
Transcriptomics, 59, 157
Transgenic, 27, 28, 161
Two-dimensional gel
electrophoresis, 157

V

visualization, 9, 41, 82, 123, 131,
135, 139, 143, 144, 149, 153, 163,
167, 168, 173

X

X!Tandem, 72, 158
xenograft, 16, 22, 23, 31, 36, 37, 38,
39, 176, 178

Printed in Singapore